Communications in Computer and Information Science   1353

More information about this series at http://www.springer.com/series/7899

Quan Yu (Ed.)

Space Information Network

5th International Conference SINC 2020
Shenzhen, China, December 19–20, 2020
Revised Selected Papers

 Springer

Editor
Quan Yu
Institute of China Electronic Equipment
Beijing, China

ISSN 1865-0929   ISSN 1865-0937 (electronic)
Communications in Computer and Information Science
ISBN 978-981-16-1966-3   ISBN 978-981-16-1967-0 (eBook)
https://doi.org/10.1007/978-981-16-1967-0

This Springer imprint is published by the registered company Springer Nature Singapore Pte Ltd.
The registered company address is: 152 Beach Road, #21-01/04 Gateway East, Singapore 189721, Singapore

Preface

This book collects the papers presented at the 5th Space Information Network Conference (SINC 2020), an annual conference organized by the Department of Information Science, National Natural Science Foundation of China. SINC is supported by the key research project of the basic theory and key technology of space information network of the National Natural Science Foundation of China, and organized by the "space information network" major research program guidance group. The aim of SINC is to explore new progress and developments in space information networks and related fields: to show the latest technology and academic achievements in space information networks, to build an academic exchange platform for researchers at home and abroad working on space information networks and related industry sectors, to share the achievements and experience of research and applications, and to discuss the new theory and new technologies of space information networks.

This year, SINC received 104 submissions, including 63 English papers and 41 Chinese papers. After a thorough review process, 13 outstanding English papers were selected for this volume (retrieved by EI), accounting for 20.6% of the total number of English papers.

The high-quality program would not have been possible without the authors who chose SINC 2020 as a venue for their publications. We are also very grateful to the Academic Committee and Organizing Committee members, who put a tremendous amount of effort into soliciting and selecting research papers with a balance of high quality, new ideas, and new applications.

We hope that you enjoy reading and benefit from the proceedings of SINC 2020.

December 2020   Quan Yu

Organization

SINC 2020 was organized by the Department of Information Science, National Natural Science Foundation of China; the Department of Information and Electronic Engineering, Chinese Academy of Engineering; the China InfoCom Media Group; and the *Journal of Communications and Information Networks*.

Organizing Committee

General Chairs

Quan Yu	Institute of China Electronic Equipment System Engineering Corporation, China
Jianya Gong	Wuhan University, China
Jianhua Lu	Tsinghua University, China

Steering Committee

Zhixin Zhou	Beijing Institute of Remote Sensing Information, China
Hsiao-Hwa Chen	National Cheng Kung University, Taiwan, China
George K. Karagiannidis	Aristotle University of Thessaloniki, Greece
Xiaohu You	Southeast University, China
Dongjin Wang	University of Science and Technology of China, China
Jun Zhang	Beihang University, China
Haitao Wu	Chinese Academy of Sciences, China
Jianwei Liu	Beihang University, China
Zhaotian Zhang	National Nature Science Foundation of China, China
Xiaoyun Xiong	National Nature Science Foundation of China, China
Zhaohui Son	National Nature Science Foundation of China, China
Ning Ge	Tsinghua University, China
Feng Liu	Beihang University, China
Mi Wang	Wuhan University, China
ChangWen Chen	The State University of New York at Buffalo, USA
Ronghong Jin	Shanghai Jiao Tong University, China

Technical Program Committee

Jian Yan	Tsinghua University, China
Min Sheng	Xidian University, China
Junfeng Wang	Sichuan University, China
Depeng Jin	Tsinghua University, China

Hongyan Li	Xidian University, China
Qinyu Zhang	Harbin Institute of Technology, China
Qingyang Song	Northeastern University, China
Lixiang Liu	Chinese Academy of Sciences, China
Weidong Wang	Beijing University of Posts and Telecommunications, China
Chundong She	Beijing University of Posts and Telecommunications, China
Zhihua Yang	Harbin Institute of Technology, China
Minjian Zhao	Zhejiang University, China
Yong Ren	Tsinghua University, China
Yingkui Gong	University of Chinese Academy of Sciences, China
Xianbin Cao	Beihang University, China
Chengsheng Pan	Dalian University, China
Shuyuan Yang	Xidian University, China
Xiaoming Tao	Tsinghua University, China

Organizing Committee

Chunhong Pan	Chinese Academy of Sciences, China
Yafeng Zhan	Tsinghua University, China
Liuguo Yin	Tsinghua University, China
Jinho Choi	Gwangju Institute of Science and Technology, South Korea
Yuguang Fang	University of Florida, USA
Lajos Hanzo	University of Southampton, UK
Jianhua He	Aston University, UK
Y. Thomas Hou	Virginia Polytechnic Institute and State University, USA
Ahmed Kamal	Iowa State University, USA
Nei Kato	Tohoku University, Japan
Geoffrey Ye Li	Georgia Institute of Technology, USA
Jiandong Li	Xidian University, China
Shaoqian Li	University of Electronic Science and Technology of China, China
Jianfeng Ma	Xidian University, China
Xiao Ma	Sun Yat-sen University, China
Shiwen Mao	Auburn University, USA
Luoming Meng	Beijing University of Posts and Telecommunications, China
Joseph Mitola	Stevens Institute of Technology, USA
Sherman Shen	University of Waterloo, Canada
Zhongxiang Shen	Nanyang Technological University, Singapore
William Shieh	University of Melbourne, Australia
Meixia Tao	Shanghai Jiao Tong University, China
Xinbing Wang	Shanghai Jiao Tong University, China

Feng Wu	University of Science and Technology of China, China
Jianping Wu	Tsinghua University, China
Xiang-Gen Xia	University of Delaware, USA
Hongke Zhang	Beijing Jiaotong University, China
Youping Zhao	Beijing Jiaotong University, China
Hongbo Zhu	Nanjing University of Posts and Telecommunications, China
Weiping Zhu	Concordia University, Canada
Lin Bai	Beihang University, China
Shaohua Yu	FiberHome Technologies Group, China
Honggang Zhang	Zhejiang University, China
Shaoqiu Xiao	University of Electronic Science and Technology of China, China

Contents

Research on Interference from 5G System to NGSO Satellite Constellation Based on K-means Clustering

Linghui Li[1], Wei Li[2], Zixuan Ren[1], Jin Jin[1,3(✉)], and Linling Kuang[1,3]

[1] Tsinghua University, Beijing, China
jinjin_sat@tsinghua.edu.cn
[2] State Radio Monitoring Center, Beijing, China
[3] Beijing National Research Center for Information Science and Technology, Beijing, China

Abstract. In the co-frequency coexistence problem between the ground 5G system and NGSO (non-geostationary Orbit) constellation system, the interference from the 5G system to NGSO satellites is a typical scenario. Due to the massive number of satellites in the NGSO satellite constellation system, the location, beam direction and beam coverage of satellites are constantly changing. There are issues that the interference calculation amount is large and the actual distribution of the 5G system is difficult to obtain. To address these issues, we analyze the system model and interference principle, put forward the method of 5G system radiation energy to reduce the calculation amount, and present the location analysis method of 5G system based on K-means clustering to reflect the actual distribution of 5G system. Based on this, the interference from the Taiyuan 5G system to the O3b system satellite constellation is simulated. Compared with the existing simulation methods, the proposed method has less computation and is more in line with the actual distribution characteristics of the specific urban 5G system.

Keywords: NGSO satellite constellation · 5G · K-means clustering algorithm

1 Introduction

In recent years, satellite communication technology has become increasingly mature and advanced, and satellite system construction is in the ascendant. SpaceX, OneWeb, Telesat, Amazon, and other companies have proposed a low-orbit communication constellation plan to build a space broadband communication system composed of NGSO satellite constellations. At the same time, terrestrial wireless mobile communication technology is also developing rapidly, and 5G communication technology has reached the commercial level [1]. In recent years, the terrestrial 5G system (IMT-2020) has gradually

This work is supported by the National Key Research and Development Program of China (2020YFB1804800), National Nature Science Foundation of China (Grant No. 91738101) and Shanghai Municipal Science and Technology Major Project (Grant No. 2018SHZDZX04) and BNRist.

Q. Yu (Ed.): SINC 2020, CCIS 1353, pp. 1–17, 2021.
https://doi.org/10.1007/978-981-16-1967-0_1

tapped the use potential of high-frequency bands in order to effectively increase the data transmission rate and system capacity. The 3GPP (3rd Generation Partnership Project) standard divides the FR (Frequency Range) of 5G NR (New Radio) into two ranges: FR1 and FR2. In the Standard Release 16, the frequency range of FR1 is 410 MHz–7125 MHz, and the frequency range of FR2 is 24.25 GHz–52.6 GHz. In the 5G system, the FR2 frequency band overlaps with the frequency bands such as Ka and Q/V commonly used in satellite communications. The FR2 frequency band division listed in the Release16 standard is shown in Table 1 [2].

Table 1. NR operating bands in FR2.

NR operating band	Uplink (UL) and Downlink (DL) operating band BS transmit/receive UE transmit/receive FUL,low – FUL,high $F_{DL,low} - F_{DL,high}$	Duplex mode
n257	26.5 GHz–29.5 GHz	TDD
n258	24.25 GHz–27.5 GHz	TDD
n259	39.5 GHz–43.5 GHz	TDD
n260	37 GHz–40 GHz	TDD
n261	27.5 GHz–28.35 GHz	TDD

At present, Ka and Q/V high-frequency bands have been widely used in satellite system services. There are many studies on interference between satellite systems [3–9] while few on interference between 5G systems and NGSO systems.

According to the main frequency band division of typical NGSO satellite constellation systems, the frequency bands overlapping with the 5G system are shown in Table 2 [10–15].

According to the frequency usage of Table 1 and Table 2, there is potential interference between NGSO satellite constellation system and 5G system. Therefore, it is necessary to study the interference between NGSO satellite constellation system and 5G system. In the interference scenario of 5G system interferes with the NGSO satellite system, due to the large beam coverage of satellite, there are many 5G system base stations and users that cause interference to the satellite, so it is often analyzed by equivalent methods [16–18]. The existing simulation analysis model mainly calculates the interference from 5G system to the single GSO (Geostationary Orbit) satellite In reference [16], the aggregate interference from the local area is calculated. Then based on the scale factor which is composed of area ratio, urban area factor and hot spot factor, the aggregate interference from all 5G systems within the coverage area of the satellite spot beam to a satellite could be obtained. Literature [17] and literature [18] introduce the concept of the central station. At the location of different central station, simulate the interference from a few base stations and terminals to satellite. The sum of the simulation

Table 2. The main frequency band where the NGSO satellite constellation system overlaps with the 5G system.

Constellation system	Link direction	Feeder link	User link
O3b system	Uplink	27.5–30.0 GHz	27.5–30.0 GHz
	Downlink	17.7–18.6 GHz 18.8–20.2 GHz	17.7–18.6 GHz 18.8–20.2 GHz
OneWeb system	Uplink	27.6–29.1 GHz 29.5–30 GHz	14.0–14.5 GHz
	Downlink	17.8–18.575 GHz 18.8–19.625 GHz	10.7–12.7 GHz
Starlink system	Uplink	27.6–29.1 GHz 29.5–30 GHz	14.0–14.5 GHz
	Downlink	17.8–18.55 GHz 18.8–19.3 GHz	10.7–12.7 GHz
Telesat system	Uplink	27.5–29.1 GHz 29.5–30.0 GHz	27.5–29.1 GHz 29.5–30.0 GHz
	Downlink	17.8–18.6 GHz 18.8–19.3 GHz 19.7–20.2 GHz	17.8–18.6 GHz 18.8–19.3 GHz 19.7–20.2 GHz
Amazon Kuiper system	Uplink	27.5–30.0 GHz	27.5–30.0 GHz
	Downlink	17.7–18.6 GHz 18.8–20.2 GHz	17.7–18.6 GHz 18.8–20.2 GHz

results is equivalent to the interference from all the base stations and terminals within the satellite beam coverage. The multiple gains are added to the base station's transmit power to reduce simulation time. There are two main problems in the current research. One is the increased amount of calculation of interference from 5G system to NGSO system satellites the NGSO satellite due to multiple characteristics of NGSO satellite constellation system, in which there is a larger number of satellites, multiple connection from earth station to satellites, constantly changing position and beam direction of the satellites compared with GSO system. The other one is that, considering the different distribution densities of 5G system worldwide, the interference is different. However, the current research does not combine the actual 5G system density, resulting in the same interference in areas with different 5G system distribution densities, which is unrealistic.

For the first problem, this paper proposes a method to obtain the EIRP (effective isotropic radiated power) of the base station and the user by fitting the CDF curve of the EIRP, which avoids the scheduling of the base station and the user, thereby reducing the amount of calculation. For the second problem, this paper proposes an equivalent method based on the K-means clustering algorithm. This method fits the 5G system density according to the building density, making this simulation scenario more realistic. The simulation results show that the method in this paper can be more in line with the actual interference from 5G system to the satellite, and the calculation amount is reduced due to the omission of the base station and user scheduling.

2 System Model

Consider a scenario where the uplink of the NGSO satellite constellation system and the 5G system coexist at the same frequency. As shown in Fig. 1, the NGSO satellite constellation system consists of S NGSO satellites and E earth stations. The 5G system is composed of N_B base stations and N_U users.

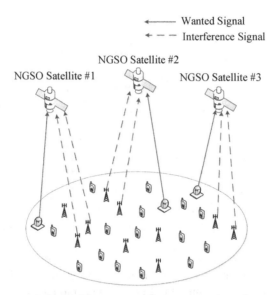

Fig. 1. NGSO system uplink and 5G system coexistence scenario at the same frequency ($S = 3$, $E = 5, N_B = 10, N_U = 15$).

When the working frequency bands of the 5G system and the NGSO satellite system overlap, the NGSO satellite system will receive the interference signal from the 5G system while receiving the useful signal of the system. The interference expression is:

$$I_{IMT} = P_{IMT} + G_{IMT}(\theta) + G_S(\phi) - L(d) \tag{1}$$

Among them, P_{IMT} is the transmission power converted from the communication bandwidth of the 5G system transmitter to the NGSO satellite system, $G_{IMT}(\theta)$ is the antenna transmission gain of the 5G system transmitter in the direction θ away from the antenna axis, $G_S(\phi)$ is the receiving antenna gain at the receiving end of the satellite system in the direction ϕ away from the main axis of the antenna, and $L(d)$ is the corresponding link loss when the distance between the transmitting end of the 5G system and the receiving end of the satellite system is d.

Compare the aggregate interference I_{IMT} with the interference threshold I_{th}. If $I_{IMT} \geq I_{th}$, there is interference from 5G system to NGSO satellite system.

3 Method

3.1 Method for Determining Radiation Energy Based on Equal EIRP

EIRP is the product of the transmit power of the power amplifier and the gain of the antenna, and is defined as follows:

$$EIRP = P_T + G_T \tag{2}$$

Among them, P_T is the antenna transmit power (dB), and G_T is the antenna transmit gain in the antenna radiation direction (dB).

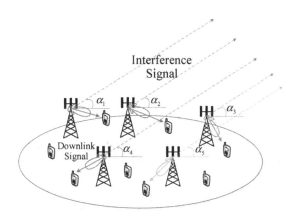

Fig. 2. Radiation of 5G system in a specific direction in space.

As shown in Fig. 2, generally, the relative azimuth between base station and user is random, the radiation energy of base station or user in each direction angle is relatively uniform, and the azimuth angle of the satellite relative to the 5G system has little effect on the EIRP of the 5G system. Since the height of the user relative to the base station is generally low, the radiation energy of base station or user at each elevation angle is not uniform. Therefore, the elevation angle of the satellite relative to the 5G system will affect the EIRP of the 5G system. Due to the large distance between satellite and ground, the elevation angles of the satellite relative to different base stations and users of 5G system are close in a local area. Therefore, the elevation angles of satellite relative to all base stations and users of 5G system in the local area $\alpha_i(i = 1, 2, ..., N_B)$ can be equivalent to the same elevation angle α. The equivalent scenario is shown in Fig. 3.

Based on the method of equal EIRP, the EIRP radiated by all 5G system base stations and users in a local area can be equivalent to the EIRP radiated by a larger base station to reduce the number of base stations and users in the local area, thereby reducing the amount of calculation. The specific expression is shown below:

$$\sum_{i=1}^{N_B} P_i G(\theta_i) + \sum_{j=1}^{N_U} P_j G(\theta_j) = EIRP \tag{3}$$

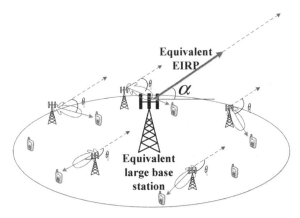

Fig. 3. Local equivalent large base station.

Among them, P_i is the transmission power converted from the communication bandwidth of the i-th 5G system transmitter to the NGSO satellite system, and $G(\theta_i)$ is the antenna transmit gain of the i-th 5G system transmitting end in a direction deviating from the antenna main axis at θ_i.

CDF (cumulative distribution function) is the integral of the probability density function, which can completely describe the probability distribution of a real random variable X. Its definition is as follows:

$$F_X(x) = P(X \leq x) \tag{4}$$

Because the satellite covers a large area and includes a large number of local areas, the location distribution and scheduling of base stations and users in each local area are different. In order to reduce the amount of calculation, the scheduling of the base station and user can be omitted by fitting the CDF curve of the 5G system base station and user EIRP. In the interval [0,1], select N_B uniformly distributed random numbers as the ordinate of the CDF curve of the base station EIRP, then the base station EIRP value is obtained from the CDF curve, and the user EIRP value is determined in the same way.

According to this method, a set of base station and user EIRP values that conform to the EIRP distribution probability can be generated, so that the scheduling of the base station and the user can be omitted, thereby simplifying the calculation.

3.2 Method for Determining the Position of Equivalent Station Based on K-means Clustering

POI (point of interest) is geospatial big data which has the advantages of large data volume, easy access, and spatial attributes, so it is easy to realize more detailed research on a large scale. Therefore, POI data has begun to be used in the research of urban commercial space in recent years, such as the identification of urban commercial centers, the characteristics and evolution of commercial agglomeration, as well as the delineation of commercial districts [19].

According to the diagram shown in Fig. 1, the NGSO satellite system receives the interference signal from the 5G system in several cities. Suppose that there are M cities within the beam coverage area of NGSO satellite #1, the analysis takes the m-th city as an example. Based on the POI data of the city, longitude and latitude, the geographic coordinates of commercial buildings in the m-th city could be obtained $\mathbf{A}_m = \{p_1, p_2, \ldots, p_a | p_i = (Lon_i, Lat_i), i = 1, 2, \ldots, a\}$. According to the locations of the buildings, the geographic coordinates could be divided into multiple disjoint subsets geographically $\mathbf{A}_m = \mathbf{A}_{m,1} \cup \mathbf{A}_{m,2} \cup \ldots \cup \mathbf{A}_{m,N_m}$.

Considering that the density of 5G base stations is positively correlated with the density of buildings, the number of buildings in the subset $|\mathbf{A}_{m,i}|$ corresponds to the number of base stations $|B_{m,i}|$ in N_m 5G hotspot areas. In order to reduce the amount of calculation, all 5G base stations and users in a hotspot area are equivalent to a large base station through EIRP equalization $\sum_{j \in \mathbf{A}_{m,j}} EIRP_j = EIRP_{m,i}$, $i = 1, 2, \ldots, N_m$.

K-means algorithm is a commonly used clustering algorithm in data mining algorithms. It groups similar elements into a group and keeps dissimilar ones as far away as possible. This algorithm has the characteristics of simple principle and better clustering effect [20]. This paper uses the K-means clustering algorithm to divide urban commercial buildings into multiple clusters to reflect the distribution of 5G system hot spots. The specific algorithm flow is as follows:

Algorithm: K-means clustering based equivalent large base station location determination algorithm

---

Step 1: Randomly select Q initial cluster centers from the original data of urban commercial buildings $\mathbf{A}_m = \{p_1, p_2, \ldots, p_a | p_i = (Lon_i, Lat_i), i = 1, 2, \ldots, a\}$.

Step 2: Calculate the Euclidean distance from the latitude and longitude coordinates of each remaining building to the Q initial cluster centers.

Step 3: Assign the remaining building points to the cluster with the smallest distance to obtain a partition set of Q clusters $(C_1, C_2, \ldots, C_Q)$.

Step 4: Calculate the centroids in each cluster $\mu_i, i = 1, 2, \ldots Q$ and compare whether the centroids and cluster centers overlap. If they do not coincide, use the centroids as the new cluster centers; otherwise, go to step 6.

Step 5: Traverse the distance between all building points and the cluster center to find a new centroid. Compare whether the centroids and cluster centers are coincident, if they do not coincide, use the centroids as the new cluster centers.

Step 6: Return to step 5 to perform the iterative operation until the cluster center and the centroid coincide, and finally the cluster division and the cluster center are obtained. The cluster center position is the equivalent large base station position of the 5G system hot spot area.

This paper uses Euclidean distance as the judgment condition. The cluster is divided into $(C_1, C_2, ..., C_Q)$. The minimum error of cluster division is:

$$E = \sum_{i=1}^{Q} \sum_{j \in C_i} \| p_j - \mu_i \| \tag{5}$$

Where Q represents the number of clusters, p_j represents the latitude and longitude coordinates of the j-th building, and μ_i represents the centroid of the i-th cluster, which is expressed as follows:

$$\mu_i = \frac{1}{|C_i|} \sum_{j \in C_i} p_j \tag{6}$$

According to the number of buildings contained in the cluster $|\mathbf{A}_{m,i}|$ and the mapping relationship, the number of interfering base stations in the cluster $B_{m,i} = f(|\mathbf{A}_{m,i}|) = k(|\mathbf{A}_{m,i}|), i = 1, 2, ..., N_m, k \text{ is constant}$ is obtained Assuming that the number of users $U_{m,i}$ is proportional to the number of base stations $U_{m,i} = kB_{m,i}, i = 1, 2, ..., N_m, , k \text{ is constant}$, the EIRP of the equivalent large base station can be obtained by using the number of base stations $B_{m,i}$ and the number of users $U_{m,i}$ in the cluster, which is:

$$\sum_{j=1}^{B_{m,i}} P_{B,j} G(\theta_{B,j}) + \sum_{k=1}^{U_{m,i}} P_{U,k} G(\theta_{U,k}) = EIRP_{m,i} \tag{7}$$

Among them, $P_{B,j}$ is the transmission power converted from the communication bandwidth of the i-th base station transmitter to NGSO satellite system, $G(\theta_{B,j})$ is the antenna transmit gain of the i-th base station transmitting end in a direction deviating from the antenna main axis at $\theta_{B,j}$, $P_{U,k}$ is the transmission power converted from the communication bandwidth of the i-th user transmitter to NGSO satellite system, and $G(\theta_{U,k})$ is the antenna transmit gain of the i-th user transmitting end in a direction deviating from the antenna main axis at $\theta_{U,k}$.

In order to simplify the calculation, the EIRP values of the base stations and users in each cluster are randomly read on the CDF curve of EIRP, based on the elevation angle of the satellite relative to the equivalent large base station.

The aggregate EIRP of all base stations and users is the EIRP of the equivalent large base station.

Using the effective isotropic radiated power $EIRP_{m,i}$ of the i-th equivalent large base station in the m-th city, the interference $I_{IMT,m,i}$ of the equivalent large base station to the satellite can be calculated as follows:

$$I_{IMT,m,i} = EIRP_{m,i} + G_S(\phi_{m,i}) - L(d_{m,i}) \tag{8}$$

Among them, $EIRP_{m,i}$ is the effective isotropic radiation power of the i-th equivalent large base station in the m-th city, $G_S(\phi_{m,i})$ is the receiving antenna gain at the receiving end of the satellite system in the direction $\phi_{m,i}$ away from the main axis of the antenna, and $L(d_{m,i})$ is the corresponding link loss when the distance between the transmitting end and the receiving end is $d_{m,i}$.

The scene diagram of the interference from the equivalent large base station group to NGSO satellite is shown in Fig. 4.

The beamwidth of the NGSO satellite is φ. There are M cities' equivalent base stations located within the satellite beam coverage, and there are N_m equivalent large base stations in the m-th city. Therefore, the aggregate interference from all equivalent base stations to satellite is:

$$I_{IMT} = \sum_{m=1}^{M} \sum_{i=1}^{N_m} I_{IMT,m,i} \tag{9}$$

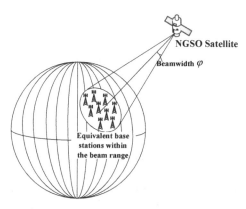

Fig. 4. Aggregate interference from equivalent large base stations to satellite.

Introduce an NGSO satellite to the NGSO satellite constellation, which should consider the interference from equivalent large base stations to S NGSO satellites of NGSO satellite constellation system. As shown in Fig. 5, assuming that the earth station is connected to S satellites, the equivalent large base stations within the coverage of S satellite beams cause interference to S satellites. Therefore, the EIRP radiated by the equivalent large base stations is:

$$\mathbf{EIRP} = [EIRP_j]_{S \times 1} \tag{10}$$

$$EIRP_j = \sum_i P_i G(\theta_{i,j}) \tag{11}$$

Among them, P_i is the transmission power converted from the communication bandwidth of the i-th transmitter of the equivalent large base station to the NGSO satellite system, and $G(\theta_{i,j})$ is the antenna transmit gain of the i-th transmitter of the equivalent large base station in the direction of the j-th satellite deviating from the antenna main axis at $\theta_{i,j}$.

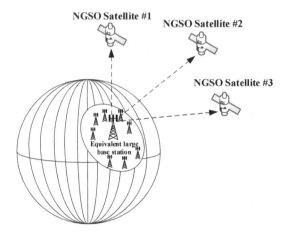

Fig. 5. The equivalent large base stations interfere with the satellite constellation ($S = 3$).

The interference from equivalent large base stations to satellite constellation is:

$$\mathbf{I} = [I_j]_{S \times 1} \tag{12}$$

$$I_j = EIRP_j + G_S(\phi_j) - L(d_j) \tag{13}$$

Among them, $EIRP_j$ is the effective isotropic radiation power of the equivalent large base station in the j-th city, $G_S(\phi_j)$ is the receiving antenna gain at the receiving end of the satellite j in the direction ϕ_j away from the main axis of the antenna, and $L(d_j)$ is the corresponding link loss when the distance between the transmitting end, and the receiving end of satellite j is d_j.

Finally, the sum of the interference from all equivalent large base stations under the beam coverage area of the satellite system is the total interference. Through the above algorithm, the location distribution of the equivalent large base station is close to the actual 5G hotspot location distribution. In addition, the scheduling between base station and user is replaced by CDF curve, which simplifies the interference calculation.

4   Simulation and Analysis

According to the actual parameter modeling of the O3b system, the O3b constellation system includes zero-inclination orbit and inclined orbit satellites. The orbit configuration of the simulated constellation is shown in Table 3.

In the heterogeneous network topology of the 5G system, macro cells generally work in low-frequency bands to meet coverage needs, and microcells work in high-frequency bands to increase system capacity. Typical application scenarios include urban microcells and indoor scenarios [21]. Considering that the 5G system and NGSO system interference frequency bands are high-frequency bands and the indoor scene penetration loss is relatively large, this article only considers the interference of the 5G system's urban microcellular application scenarios to the O3b system satellite constellation. The

Table 3. O3b system track configuration.

Number of track surfaces	Number of satellites/orbital surface	The height of the track	Inclination of orbit
1	60	8062 km	0°
6	24	8062 km	40°
6	24	8062 km	70°
6	24	1400 km	10°
1	60	9000 km	0°
6	24	9000 km	40°
6	24	9000 km	70°
1	60	8062 km	0°
6	24	8062 km	40°

network topology is shown in Fig. 6. Each macrocell contains 3 identical hexagonal sectors, and the distance between the macrocells is 200 m [21]. A micro base station is deployed within each macro cell and the distance between micro base stations is not less than 50 m. Each micro base station schedules 3 users.

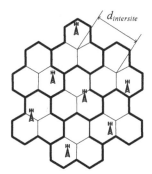

Fig. 6. 5G system microcellular network topology.

There are generally one or more cities under the coverage of O3b satellite beams. Since the aggregate interference of multiple cities to satellites is the linear sum of the interference of different cities, this article uses a city (Taiyuan City) as an example to simulate the interference from 5G system to O3b satellite constellation.

According to the method proposed in this article, using the commercial POI data of Taiyuan in 2020 [22], 1,840 pieces of commercial information were extracted. According to the clustering method, the centers of multiple clusters are obtained, which are equivalent to the positions of the equivalent large base stations. As shown in Fig. 7, the position of the equivalent large base station is represented by a blue solid dot, and the center of Taiyuan (37.86°N, 112.58°E) is represented by a triangle.

According to the network topology of the urban micro-cell of the 5G system, there are approximately 30 micro-base stations distributed in the urban area of 1 km^2. According to the area of 6909 km^2 in Taiyuan [23], set the proportion of urban area to 7%, the

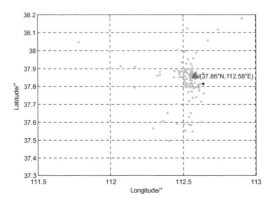

Fig. 7. Location of equivalent large base station. (Color figure online)

proportion of hot spot to 100%, and the ratio of active base stations to 20%. After the calculation, the total number of micro base stations in Taiyuan is about 2902. Combined with the total number of POI data of 1840, the proportional approximate relationship between the number of interfering base stations $B_{m,i}$ and the number of buildings $|\mathbf{A}_{m,i}|$ in the i-th cluster in the m-th city is obtained $B_{m,i} = k\left(|\mathbf{A}_{m,i}|\right), k = 1.58$.

Simulate the scheduling of base stations and users in a local area of 2 km², with an interval of 1°, obtain the EIRP values of the base stations and users at an elevation angle of 0° to 90°, and fit the CDF curve. Figure 8 shows the actual simulation curve and CDF fitting curve of the EIRP of the base stations at different elevation angles. Figure 9 shows the actual simulation curve and CDF fitting curve of the EIRP of the users at different elevation angles. It can be seen that the fitted curve is consistent with the real curve. As the elevation angle increases, the EIRP of the base stations and the users gradually decreases.

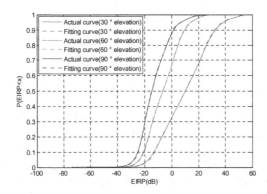

Fig. 8. The actual simulation curve and CDF fitting curve of the EIRP of base station.

Randomly read the EIRP values of the base stations and the corresponding users on the fitted CDF curves of the EIRP of the base stations and users, and then obtain the EIRP values of the equivalent large base stations corresponding to the clusters. Add the

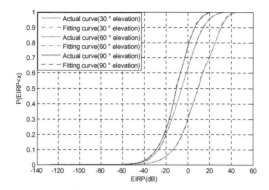

Fig. 9. The actual simulation curve and CDF fitting curve of the EIRP of user.

interference from each equivalent large base station to the satellite, and finally get the aggregate interference from all the equivalent large base stations to the satellite.

Set up an earth station in the center of Taiyuan (37.86°N, 112.58°E), which can access 5 O3b satellites at the same time. Considering that the 5G system works in TDD mode, the base station/terminal performs signal transmission and reception operations in the same frequency band. To simulate the mobility of satellites and users, the Monte Carlo simulation is used to complete the corresponding system-level simulation. The EIRP of the equivalent large base station of each snapshot is different, the satellite position is different, and enough snapshots are simulated, so that the simulation result is closer to the actual system.

In order to verify the correctness of the proposed method, two comparison methods are used. The specific methods are as follows:

Comparison method 1 [16]: First of all, calculate the aggregate interference I_1 (dB) from the 5G system in a local area of 2 km^2 in the center of Taiyuan (37.86°N, 112.58°E). Then, according to the area ratio D of the total area of Taiyuan to the local area, the urban factor Ra, and the hot spot factor Rb, the aggregate interference I_t (dB) from all 5G systems in Taiyuan to satellites is calculated.

$$I_t = I_1 + 10\log_{10}(D \times R_a \times R_b) \tag{14}$$

Comparison method 2 [18]: Adopt the concept of the central station, and simulate the interference from a few base stations and users to satellites at random locations in Taiyuan City. The local area is approximately 2 km^2. Accumulate the interference of each simulation until the number of simulated base stations and users reaches the number that should be within the scope of Taiyuan City, that is, 3000 micro base stations and 9,000 users.

The specific simulation parameters are shown in Table 4.

Table 4. Parameters of interference simulation.

System	Simulation parameters	Value
5G system	Frequency/GHz	29
	Working bandwidth/MHz	200
	Total area of Taiyuan City/km^2	6909
	Area of local area/km^2	2
	Number of base station scheduled users	3
	Base station height/m	10
	User height/m	1.5
	Downtilt angle of Base station	10°
	Urban factor Ra	0.07
	Hot spot factor Rb	1
	Network loading factor	20%
	Base station TDD activity factor	100%
	IMT Antenna pattern	Rec ITU-R M.2101
	Maximum base station terminal output power/(dBm 200MHz^{-1})	28
	Maximum user terminal output power/(dBm 200 MHz^{-1})	22
	Element gain/dBi	8
	Base station antenna array configuration (Row × Column)	16 × 16 elements
	User antenna array configuration (Row × Column)	4 × 4 elements
	Horizontal/vertical 3 dB beamwidth of single element	65°
	Horizontal/vertical front-to-back ratio	30 dB
	Horizontal/Vertical radiating element spacing	0.5 of wavelength for both H/V
	Array Ohmic loss	3 dB
	User body loss	4 dB
	Noise figure	12 dB
	Conducted power per antenna element/(dBm 200 MHz^{-1})	10
Satellite system	Frequency /GHz	29
	Working bandwidth /MHz	250

(*continued*)

Table 4. (*continued*)

System	Simulation parameters	Value
	Space station peak transmit antenna gain/dBi	45
	Space station diameter/m	0.93
	Space station antenna mode	REC-1528
	Space station system noise temperature/K	600
	Latitude and longitude of earth station	(37.86°N, 112.58°E)
	Earth station height/m	10
	Earth station minimum elevation angle/°	30
	Number of satellites connected to earth station	5

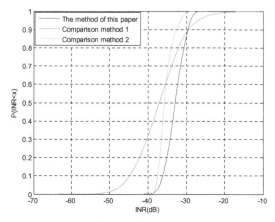

Fig. 10. Cumulative distribution curve of interference from 5G system to satellite constellation. (Color figure online)

Figure 10 shows the interference simulation results of the three methods. The cumulative distribution curve of the interference value of this method is represented by a blue line, the cumulative distribution curve of the interference value of the comparison method 1 is represented by a red line, and the cumulative distribution curve of the interference value of the comparison method 2 is represented by a purple line. The uneven distribution of the 5G system will cause the cumulative distribution curve to approach the blue and purple lines, which is in line with the actual situation. The uniform modeling method will cause the cumulative distribution curve to approach the red line, which doesn't match the actual characteristics.

The method proposed in this paper can better obtain the real position of the 5G system's concentration center, thereby simulating the real EIRP spatial distribution within the satellite beam coverage. The result of interference analysis is more in line with the actual situation.

In addition, the simulation duration of the three methods is shown in Table 5. The simulation duration of the method proposed in this paper is similar to that of comparison method 1, and the simulation duration is greatly reduced compared to comparison method 2. This is because the method proposed in this article does not need to simulate the scheduling between the base station and the user in each snapshot but only needs to randomly select a value on the fitted CDF curve, thereby reducing the simulation duration.

Table 5. Simulation duration comparison of the three methods.

Method	Duration
Clustering method	7662 s
Comparison method 1	5023 s
Comparison method 2	1763631 s

5   Conclusion

In the 5G system's interference scenario to the NGSO system satellite constellation, due to a large number of satellites, the continuous change of satellite positions, and the many 5G system base stations and users within the satellite beam coverage, there are several issues that the amount of interference calculation is large, and it is difficult to obtain the actual distribution of the 5G system. Based on the method of EIRP equalization, this paper analyzes the method that the EIRP of base stations and users in 5G hotspots is equivalent to the EIRP of a large base station. According to the commercial agglomeration characteristics of cities in the satellite coverage area, the K-means clustering method is used to generate clusters of urban commercial buildings. Map the clusters to the location and scale of the 5G hotspot area in the city to determine the location and interference of the equivalent large base station. Finally, by summarizing the interference from all equivalent large base stations, the interference from the 5G system with the satellite beam coverage is obtained. The simulation results show that the simulation results of the method proposed in this paper can be more in line with the actual 5G system distribution and reduce the simulation time.

References

1. Lotus Magazine, China officially enters the 5G commercial era. https://www.sohu.com/a/376 040576_556660
2. 3GPP TS 38.104.: Base Station (BS) radio transmission and reception (Release 16). ETSI (2020)

3. Liu, Q., Ge, X., et al.: Overview of the global situation of frequency resources of non-geostationary orbit broadband communication constellations. https://mp.weixin.qq.com/s/XeHf3ppub9mcelXC1EO2Zg

4. Li, T., Jin, J., Yan, J., et al.: Resource allocation in NGSO satellite constellation network constrained by interference protection to GEO satellite network. In: 70th International Astronautical Congress, Washington D.C., United States, IAC-19-B2.3.13 (2019)

5. Lin, Z.Q., Jin, J., Yan, J., Kuang, L.L.: A method for calculating the probability distribution of interference involving mega-constellations. In: 71th International Astronautical Congress, IAC-20-B2.1.3 (2020)

6. Jin, J., Li, Y.Q., Zhang, C., et al.: Occurrence probability of co-frequency interference and system availability of non-geostationary satellite system in global dynamic scene. J. Tsinghua Univ. (Sci. Tech.) **58**(009), 833–840 (2018)

7. Zhang, C., Jiang, C.X., Jin, J., et al.: Spectrum sensing and recognition in satellite systems. IEEE Trans. Veh. Technol. **68**, 2502–2516 (2019)

8. Zhang, C., Jin, J., Zhang, H., et al.: Spectral coexistence between LEO and GEO satellites by optimizing direction normal of phased array antennas. China Commun. **15**(006), 18–27 (2018)

9. Zhang, C., Jin, J., Kuang, L.L., et al.: Blind spot of spectrum awareness techniques in non-geostationary satellite systems. IEEE Trans. Aerosp. Electron. Syst. **54**, 3150–3159 (2018)

10. ITU, O3B-C Satellite network in IFIC 2854. https://www.itu.int/sns/demowic20.html

11. ITU, L5 Satellite network in IFIC 2910. https://www.itu.int/sns/demowic20.html

12. ITU, CANSAT-LEO Satellite network in IFIC 2850. https://www.itu.int/sns/demowic20.html

13. ITU, STEAM-1 Satellite network in IFIC 2884. https://www.itu.int/sns/demowic20.html

14. ITU, STEAM-2 Satellite network in IFIC 2884. https://www.itu.int/sns/demowic20.html

15. ITU, USASAT-NGSO-8A Satellite network in IFIC 2916. https://www.itu.int/sns/demowic20.html

16. Wei, M.: Research on coexistence of interference between 5G system and fixed satellite service. Beijing Univ. Posts Telecommun **2019**, 1–13 (2019)

17. Deng, Z.L., Pan, Z., Wang, T., et al.: Simulation analysis of electromagnetic compatibility between 5G base station and satellite based on Visualyse. China Radio **2018**(10), 59–63 (2018)

18. Zhang, G.L.: Research on coexistence of interference from IMT-2020 system to fixed satellite services. Beijing Univ. Posts Telecommun. **2018**, 1–49 (2018)

19. Lin, Q., Sun, F., Wang, X.M., et al.: Research on the hierarchical system of Beijing's commercial center based on POI data. J. Beijing Normal Univ. (Nat. Sci.) **55**(03), 415–424 (2019)

20. Luo, J.F., Hong, D.D.: A K-means clustering algorithm based on density and distance. Softw. Eng. **23**(10), 23–25+4 (2020)

21. Han, R., Zhang, L., Li, W., et al.: Analysis of Interference from IMT-2020 (5G) system to satellite broadcasting system in the frequency band of 24.65~25.25 GHz. Telecommun. Sci. **34**(07), 108–115 (2018)

22. Amap, AMAP Inside. https://console.amap.com

23. Administrative Division Network, Taiyuan City Overview Map. https://www.xzqh.org/html/show/sx/3524.html

Fast Calculation of the Probability Distribution of Interference Involving Multiple Mega-Constellations

Ziqiao Lin[1], Jin Jin[2,3(✉)], Jian Yan[2,3], and Linling Kuang[2,3]

[1] Department of Electronic Engineering, Tsinghua University, Haidian, Beijing, China
[2] Beijing National Research Center for Information Science and Technology, Haidian, Beijing, China
jinjin_sat@tsinghua.edu.cn
[3] Tsinghua University, Haidian, Beijing, China

Abstract. Recently, mega-constellation plans, such as Starlink and OneWeb, have been proposed successively. Mega-constellations compose of far more satellites than traditional constellations, with a variety of satellites of different orbital altitudes and complex constellation configurations. A massive number of satellites with different orbital configurations will lead to a significant increase in the computation of the interference probability distribution, and even the failure of common PC to complete the simulation. Based on this, we propose a fast method to calculate the interference probability distribution of mega-constellations, which is especially suitable for calculating satellites with different orbital altitudes and configurations. By dividing the visible airspace of the earth station into sub-airspaces according to probability, the interference from dynamically moving satellites is characterized as the interference caused by static satellites in different sub-airspaces. The interference probability distribution is calculated by combining the occurrence probability of satellites with different orbital altitudes in the sub-airspace. The simulation results indicate that the proposed method greatly improves the computational efficiency and keeps almost the same accuracy as the traditional method. The proposed method can be used to rapidly compute the interference probability distribution of mega-constellations involving multiple orbital altitudes and configurations.

Keywords: Mega-constellation · The probability distribution of interference · Satellite occurrence probability · Fast calculation

1 Introduction

In recent years, a growing number of mega-constellation plans have been put forward: the Telesat constellation is scheduled to launch 1671 NGSO (non-geostationary orbit) satellites [1]; Amazon's Kuiper constellation has been granted the use of Ka frequency band with a total of 3236 satellites [2]; Starlink constellation, operated by SpaceX, proposes a "Gen2 system" consisting of 30000 low earth orbit satellites, with the 11943 satellites

© Springer Nature Singapore Pte Ltd. 2021
Q. Yu (Ed.): SINC 2020, CCIS 1353, pp. 18–34, 2021.
https://doi.org/10.1007/978-981-16-1967-0_2

previously planned for launch, which will bring the total quantity of satellites to 41943 [3, 4] and there are over 700 satellites in orbit currently [5]; OneWeb plans to increase satellites from 716 to 47844 in Phase II [6]. Moreover, mega-constellations generally possess more satellites than traditional constellations and the constellation configuration is more complex, usually including multiple sub-constellations with different orbital altitudes. With the increased quantity of satellites, the problem of inter-satellite co-frequency interference becomes more prominent [7], and corresponding interference mitigation strategies should be adopted to avoid the interference with other constellation systems.

The probability distribution of interference is a common evaluation indicator to describe the inter-constellation interference. It is the premise to analyze interference and formulate mitigation strategies. Presently, the methods to calculate the probability distribution of interference mainly include the extrapolation method and the numerical method. The extrapolation method is usually procured by extrapolating the satellite orbital position and compile statistics the occurrence time proportion of different interference values. To acquire an accurate interference probability distribution, the extrapolation method needs to be simulated for a long time to ensure that all possible satellite locations are covered, and the simulation duration is generally several months. Numerical methods have been less studied. In [8, 9], the probability distribution of interference is calculated in the global integral by deriving the satellite probability density function, and the computational complexity is reduced through the adaptive step and Gauss quadrature rule, but no specific analysis is given on the implementation details. In [10], the concept of the reference satellite setting region is proposed, which generates different constellation snapshots by placing the reference satellites at different locations within the setting region. The interference between different snapshots is calculated cyclically to obtain the final interference probability distribution. As the number of constellation layers increases, the computation will increase multiplicatively.

Relatively, traditional constellations have fewer satellites and common PC can complete the simulation through the extrapolation method. Accordingly, most of the current researches focus on obtaining more accurate interference analysis models in different scenarios while less research on simplifying the calculation of interference probability distribution. Because of the huge quantity of mega-constellation satellites, adopting the extrapolation method will lead to a substantial increase in the computation effort, and even the situation that common PC may not be able to handle the interference simulations. Starlink specifically points out in its analysis report [11] that: "The time domain simulation requires longer run times to provide accurate results…. This can lead to prohibitively long simulation run times, given the sizes of the NGSO constellations and the number of analysis to perform." It should be explained that only 1600 satellites have been simulated in this report. Additionally, mega-constellations usually possess multiple sub-constellations with different orbital altitudes and configurations, so how to reduce the computational complexity of mega-constellations involving multiple orbital altitudes will be an issue to be considered in future research [12–16].

To address these issues, we propose a method to fast calculate the probability distribution of interference involving mega-constellations, which is suitable for calculating a variety of satellites with different orbital altitudes. The interference from different visible satellites on communication links is characterized as the interference caused by probabilistic satellites in different parts of visible airspaces, which drastically reduces the computational amount without extrapolating orbits and generating constellations. Simulation results indicate that the proposed method can be used to fast evaluate the interference involving multiple mega-constellations.

2   Method

2.1   Concept and Interference Model

We consider a spectral coexistence scenario of a traditional NGSO (non-geostationary orbit) constellation system A (with a small number of satellites) and a NGSO mega-constellation system B in the downlink case. As shown in Fig. 1, the constellation A is the interfered-with system, including an accessing satellite (the interfered-with satellite) and an earth station. Mega-constellation system B is the interfering system composed of S NGSO satellites, among which $s(s < S)$ system B satellites visible to the earth station of system A will interfere the satellite-ground communication link of system A. For any earth station, when the elevation of the accessing satellite is larger than a certain angle, the quality of the communication link will meet the system operation requirements, and this elevation is called the minimum operation elevation E_{min} of the earth station. Therefore, the airspace in which the earth station can access the satellite is the airspace with an elevation greater than E_{min}, which is defined as the visual airspace.

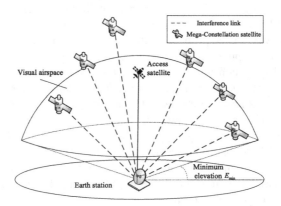

Fig. 1. Interference scenario.

We define the visual airspace as D. In this paper, we divide the visible airspace D into N subsets based on a certain strategy. Each subset represents a smaller airspace, denoted as D_i, $i = 1, ..., N$. We refer to D_i as *sub-airspace*, which satisfies $D_i \subset D$ and $D_i \neq \varnothing$. For each sub-airspace D_i, the occurrence probability of the constellation system satellite in D_i is called the *constellation satellite occurrence probability*, denoted as $p_{c,i}$. The method in this paper will be elaborated on the above basis, which is mainly consists of three parts:

1. Calculation of the constellation satellite occurrence probability $p_{c,i}$ of each sub-airspace;
2. Division of sub-airspaces on the basis of the constellation satellite occurrence probability $p_{c,i}$;
3. Calculation of interference from satellites in different sub-airspaces and obtain the probability distribution of interference.

2.2 Probability Calculation

The formula for calculating the occurrence probability of satellite in the visible airspace of an earth station is given in literature [17]. When the total number of satellites in the constellation is n, for a given earth station, the probability of any one satellite in the constellation occurs in a sub-airspace D_i is:

$$p = n \cdot \frac{A}{2\pi^2} \frac{1}{\sin \alpha} \frac{1}{\cos L} \tag{1}$$

where A is the area of the sub-airspace D_i (sterad), α is the angle between the ground track and the latitude line of the center of the D_i, and L is the latitude of the center of the D_i, expressed as:

$$A = \Delta\theta_\varepsilon \cdot \Delta\theta_\beta \, (rectangle) \tag{2}$$

$$A = \frac{\pi}{4} \cdot \Delta\theta_\varepsilon \cdot \Delta\theta_\beta \, (circular) \tag{3}$$

$$\alpha = \arccos \frac{\cos i}{\cos L} \tag{4}$$

$$L = \arcsin(\cos \theta_\varepsilon \cdot \sin L_0 + \sin \theta_\varepsilon \cdot \cos L_0 \cdot \cos A_z) \tag{5}$$

where $\Delta\theta_\varepsilon$ and $\Delta\theta_\beta$ are the difference in elevation and azimuth of D_i corresponding to the difference in geocentric angle respectively, i is the inclination, L_0 is the latitude of the earth station, A_z is the azimuth of the center of D_i to the earth station, and θ_ε is the geocentric angle between the center of D_i and the earth station.

The formula given in [17] is only applicable to the case that satellite constellations with the same orbital altitude. Mega-constellations generally involve different orbital altitudes, therefore, the occurrence probability of constellation satellite should be the superposition of the occurrence probability of the satellite with different orbital altitudes, which needs further calculation.

As demonstrated in Fig. 2, we take into account that a mega-constellation is divided into M sub-constellations based on different orbital altitudes, that is, there are M different orbital altitudes in mega-constellation. Suppose that the occurrence probability of satellites of M orbital altitudes in the sub-airspace D_i is $p_1, p_2, ..., p_M$ respectively. Since the occurrence probability of satellites with different orbital altitudes is independent of each other in the sub-airspace D_i, the occurrence probability of satellites p_c with different orbital altitudes can be acquired by the addition formula:

$$p_c = \sum_{l=1}^{M} p_l - \sum_{1 \le l < k \le M} p_l p_k + \sum_{1 \le l < k < m \le M} p_l p_k p_m + ... + (-1)^{n-1} p_1 ... p_M \tag{6}$$

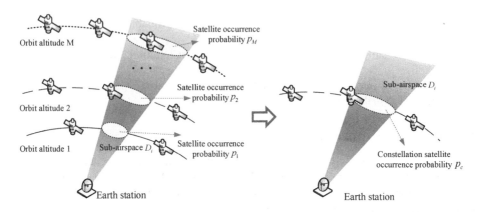

Fig. 2. Constellation satellite occurrence probability.

Herein, p_c is the occurrence probability of any satellite at any orbital altitudes within the constellation in sub-airspace D_i, we refer to as the constellation satellite occurrence probability. This formula can also be obtained by the equation:

$$p_{c,i} = 1 - \prod_{i=1}^{M} (1 - p_i) \tag{7}$$

It should be noted that the occurrence probability of satellites with different orbital altitudes in the sub-airspace D_i is independent of each other. Accordingly, the proposed method is also applicable to sub-constellations with different orbital altitudes in the same constellation and constellations with different orbital altitudes.

2.3 Visual Airspace Division

We partition the visual airspace D into several sub-airspaces. Sub-airspace division satisfies the following three criteria:

1. $\bigcup\limits_{i=1}^{N} D_i = D$;
2. $D_i \cap D_j = \varnothing, i \neq j$;
3. $p_{c,i} \approx 1$.

As the latitude increases, the distribution of satellites becomes increasingly dense. Since the different latitudes corresponding to different sub-airspace, to ensure that criterion 3 is met, the unequal area division will be adopted.

As shown in Fig. 3, the visible airspace of the earth station is divided into sub-airspaces in the geocentric dimension. Theoretically, the visual airspace can be divided by any shape, we take the square partition as an example to illustrate. According to formula (1), when the same latitude sub-airspaces with the same probability, the area of the partitioned sub-airspaces are the same. With the growth in latitude, the distribution

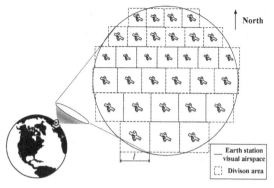

Fig. 3. Earth station visual airspace division sub-airspaces (geocentric dimension).

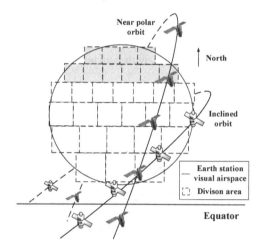

Fig. 4. The division of visual airspace involves inclined orbit constellations.

of satellites is increasingly dense, and the area of the divided airspace will gradually decrease.

As demonstrated in Fig. 4, when the satellite orbit is a polar/near-polar orbit, in probability, satellites may occur in all sub-airspaces within the visible airspace. When the satellite orbit is inclined, if the orbital inclination is low, it will be impossible for satellites to appear in some sub-airspaces, such as the grey sub-airspaces in Fig. 4. Hence, when calculating the occurrence probability of constellation satellites involving inclined orbits, it is necessary to judge the relationship between the latitude of the sub-airspace and the orbital inclination. When the latitude of sub-airspace is higher than the orbital inclination, the occurrence probability of satellite in this constellation will not be calculated.

When the satellite communication parameters are determined, the interference value is only related to different satellite positions. When the area of the sub-airspace is small, the interference caused by the interfering system satellite at any position in the sub-airspace to the interfered-with system does not change much. Base on this, a virtual satellite can be placed in the sub-airspace, and the interference caused by the virtual satellite to the interfered-with link is equivalent to the interference generated by the satellite at any position in the sub-airspace to obtain the interference value. In the mega-constellation, there are a massive number of satellites. From Eq. (1), it can be seen that if the number of satellites n is large, the area of the sub-airspace satisfying criterion 3 is small enough. In this instance, the occurrence of the satellite at any location in the sub-airspace possesses little impact on the calculation of the interference. To facilitate the calculation, the satellite is uniformly placed in the center of the sub-airspace in the following section.

2.4 Calculation Method of the Probability Distribution of Interference

2.4.1 Interference Calculation Method

Downlink interference is considered in this paper. The interference scenario is demonstrated in Fig. 5, including the interfered-with constellation A, its earth station and the interfering constellation B.

Fig. 5. Interference between constellation systems.

The interference of satellite j in constellation A interfered by i-th satellite in constellation B is as follows:

$$I_i = p'_{t,i} G_t(\theta_1) G_r(\theta_2) (\frac{\lambda}{4\pi d_i})^2 \tag{8}$$

where $G_t(\theta_1)$ is the transmit antenna gain of i-th satellite in the off-axis angle θ_1 direction (direction of the earth station) and $G_r(\theta_2)$ is the receive antenna gain of the earth station in the off-axis angle θ_2 direction (direction of i-th satellite). Additionally, λ means the wavelength, d_i is the distance from the satellite i to the earth station and p'_i donates the equivalent transmit power converted from p_i to the overlapping bandwidth. Suppose that the carrier central frequency of constellation A is the same as that of constellation B, and then the corrected transmit power can be derived by:

$$p'_{t,i} = p_{t,i} \cdot \min(\frac{W_{conA}}{W_{conB}}, 1) \tag{9}$$

where W_{conA} and W_{conB} is the communication bandwidth of constellation A and B.

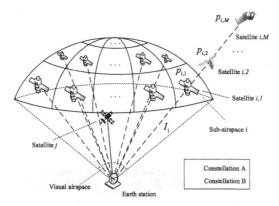

Fig. 6. The earth station is interfered by a mega-constellation.

We take into account that constellation B is a mega-constellation, as shown in Fig. 6, the visible airspace of the earth station is divided into several sub-airspaces by the method in Sect. 2.3. In each sub-airspace, the occurrence probability of constellation satellite is 1. In Fig. 6, satellite j is interfered by multiple satellites of constellation B. For interfering satellites in sub-airspace i, the occurrence probability of constellation satellite is essentially calculated from the occurrence probability of satellite with multiple orbital altitudes. Therefore, satellites with different orbital altitudes in sub-airspace i will interfere with satellite j, and the interference is the superposition of interference satellites with multiple orbital altitudes. Suppose that the occurrence probability of satellite at different orbital altitudes in sub-airspace i is $p_{i,1}, ..., p_{i,M}$ and the interference generated by it is $I_{i,1}, ..., I_{i,M}$. When the satellite communication parameters are determined, the interference value is only related to the position of the different satellites. Hence, the average interference caused by a satellite in subspace i to satellite j can be expressed as follows:

$$I_i = \sum_{l=1}^{M} I_{i,l} \cdot p_{i,l} \tag{10}$$

Similarly, since the visible airspace is divided into N sub-airspaces, all constellation B satellites in each sub-airspace interfere with the link between satellite j and the earth station. Then, the earth station is subject to aggregate interference I from constellation B is:

$$I = \sum_{i=1}^{N} I_i \tag{11}$$

Aggregate interference-to-noise ratio (I/N) is:

$$\frac{I}{N} = \frac{\sum_{i=1}^{M} \delta_i p_i' G_t(\theta_1) G_r(\theta_2)(\frac{\lambda}{4\pi d_i})^2}{K T_A W_{conA}} \tag{12}$$

where T_A is the noise temperature at the receiving end of the earth station in constellation A and K is the Boltzmann coefficient.

2.4.2 Calculation Method of Interference Probability Distribution

As described in Sect. 2.4.1, the position of the interfering constellation satellite and the communication parameters have been determined. Therefore, the interference can be determined when the interfered-with constellation satellite is determined. Based on this, the probability of the interference value is only related to the occurrence probability of the interfered-with satellite. In this paper, the method proposed in [10] is adopted to calculate the probability of the interfered-with satellite, which is specifically expressed as follows:

When the distribution of the constellation satellite is determined, the accessing satellite can be determined. Accordingly, the probability of accessing satellite is equivalent to the probability of the current satellite distribution. If the orbital parameters of one of the satellites in the constellation at a certain moment are known, in combination with the constellation configuration parameters, the distribution of all the satellites in the constellation can be deduced. The satellite distribution at a certain moment and this satellite referred to as the snapshot and the reference satellite respectively, where the reference satellite can be any one of the satellites in the constellation. Consequently, if the occurrence probability of the reference satellite is known, the probability of the snapshot can be deduced, and thus the probability of the accessing satellite can be procured.

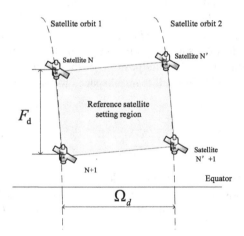

Fig. 7. The relative position of Walker constellation and setting region.

To acquire all possible access case, it is necessary to place reference satellites around the globe to generate a snapshot and get all possible snapshots. The Walker constellation is currently used by most constellations due to its good global and latitudinal coverage characteristics. According to the orbital configuration characteristics of Walker constellation, the region in which the reference satellite is placed can be narrow down to a certain region, which is called the setting region. Its scope is composed of adjacent satellites in the same orbit (phase difference in the same orbit F_d) and adjacent satellites in adjacent orbits (the right ascension of ascending node difference Ω_d), as shown in Fig. 7. The occurrence probability of the reference satellites is calculated by dividing the region within the setting region, setting up the reference satellites and obtaining the probability of all snapshots. This paper expands the statement on this basis.

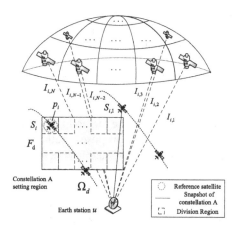

Fig. 8. Calculate the interference value and probability value involving mega-constellation.

As demonstrated in Fig. 8, K regions are divided in the interfered-with constellation A setting region and reference satellites are set. The reference satellite $S_i(i = 1, .., K)$, whose satellite occurrence probability is p_i, is selected to generate a snapshot. In this snapshot, the satellite $S_{i,1}$ is accessed, so the probability of access to the satellite $S_{i,1}$ is p_i. The visible airspace is divided into N sub-airspaces, in which satellites will interfere with the link between the satellite $S_{i,1}$ and the earth station, whose interference is $I_{i,1}, ..., I_{i,N}$ respectively. The aggregate interference I_i is calculated by formula (10) and (11), it is only related to the current position of satellite $S_{i,1}$. Therefore, the probability of interference I_i is the probability of the access satellite $S_{i,1}$, namely:

$$p(I_i) = p_i \tag{13}$$

The setting region is divided into K regions. The reference satellites in different regions will generate different snapshots, and the earth station will connect to different satellites, resulting in different aggregate interference. To acquire all the possible interference, it is imperative to calculate the interference values from K regions and obtain the interference $I_{K \times 1}$ and probability $P_{K \times 1}$. To get the probability distribution of interference, it is required to merge different interference values into corresponding interference intervals. The probabilities corresponding to different interference values will be accumulated into the probability of the corresponding interference interval, which is expressed as follows:

$$P(I) = \sum_{i=1}^{K} p(I_i), I - 1 \leq I_i < I, I_i \in I_{K \times 1}, p(I_i) \in P_{K \times 1} \tag{14}$$

where $P(I)$ is the probability that the interference value is in the interval $[I - 1, I]$. The process is repeated for all interference values and probability. Finally, we get the probability distribution of interference.

It should be noted that if the interfered-with constellation fails to satisfy the Walker constellation condition, the orbit of the interfered-with constellation still needs to be extrapolated to calculate the aggregate interference from the mega-constellation at each time step, so as to obtain the probability distribution of interference. However, as the mega-constellation can be approximated by the proposed method, the orbital extrapolation is only the traditional constellation with fewer satellites, so the computational amount will not be significantly improved.

Through this method, the dynamically changing satellites become static and the interference calculation is simplified to calculate the interference from different sub-airspaces, which can drastically reduce the computational amount involving multi-orbital altitude mega-constellations. Compared to the method proposed in [10], when the total number of satellites and layers increase, the proposed method only changes the number of sub-airspace and the interference caused by satellites in the sub-airspaces. There will not be a substantial increase in computational effort. Theoretically, the literature [10] method is applicable to all constellations, but it is more efficient for mega-constellations, while the proposed method is only appropriate for mega-constellations and can be applied to most of the current mega-constellations. Besides, the proposed method is required to divide the visible airspace according to the probability and procure a better approximation when the satellites in the visible airspace are uniformly distributed, hence, it is more suitable for constellations with relatively uniform satellite distribution.

3   Simulation and Analysis

To compare the effect of the proposed method with other methods, the parameters in Table 1, Table 2 and Table 4 are used to complete the simulation and analyze the probability distribution of downlink interference. Wherein, Table 1 is the simulation parameters of constellation system, constellation A is the interfered-with constellation, which is the traditional NGSO constellation; constellation B is an interfering mega-constellation, which is a single orbital altitude constellation. Table 4 shows the interfering mega-constellation involving multi-orbital altitudes (B_n represents the n-th sub-constellation). Walker constellation is used in the simulation (specifically, Walker constellation can be divided into star constellation and δ constellation). Table 2 shows the communication simulation parameters. Constellation A adopts the distance-prioritized access scheme, with its satellites using a dynamic spot beam with staring antenna and the antenna pattern is ITU-R S.465-5 [18]. The constellation B is the one using the fixed beam with a ground-oriented antenna and the antenna pattern is ITU-R S.1528 [19].

As illustrated in Fig. 9 and Fig. 10, it is the comparison result among the proposed method, the extrapolation method (60 days, 5 s time step) and the literature [10] method. The simulated constellation is a single orbital altitude constellation. It can be seen that the cumulative probability distribution curves of the three methods are almost the same, and there is an error of about 0.15 in the probability distribution of interference curve around -10 dB, which is caused by the static approximation to the dynamically changing

Table 1. Simulation parameters of the constellation system.

Parameter	Constellation A	Constellation B
Orbital altitude	1000 km	1200 km
Orbital inclination	88°	86°
Number of satellites	200	11250
Number of orbital planes	10	75
Phase factor	1	1
Position of the earth station	[116.388°E, 39.9289°N]	
Time step	5 s	
Minimum elevation	15°	

Table 2. Simulation parameters for communication

Parameter	Value
Maximal Tx* power of satellite (B)	7 dBW
Down-link bandwidth (A)	36 MHz
Down-link bandwidth (B)	49 MHz
Down-link carrier frequency	20.075 GHz
Peak gain of ES RX* antenna (A)	38.7 dB
Beamwidth of ES RX antenna (A)	2°
Peak gain of satellite TX antenna (B)	39.9 dB
Beamwidth of satellite TX antenna (B)	5°
Noise temperature of ES receiver	270 K

*Tx = Transmit and Rx = Receive
*ES = Earth station

satellite. Figure 11 illustrates the sub-airspaces segmentation, the corresponding satellites and probability (elevation/azimuth dimension), where the red solid line represents the lowest elevation angle and the yellow dot represents the satellite position. Because the satellite position is limited in different sub-airspaces, the satellite distribution is relatively uniform, which leads to some errors, but it has little influence on the cumulative disturbance distribution curve.

Besides, although the sub-airspace area is the same with the same probability and latitude in the geocentric dimension, the sub-airspace area for the low-elevation is smaller in the elevation/azimuth dimension. It will gradually increase as the elevation angle rises, which is consistent with the results deduced in [20].

The simulation duration of the three methods is shown in Table 3. It can be perceived that the efficiency of the proposed method is extremely improved compared with the extrapolation and the literature [10] method. It should be noted that the PC used in this simulation can only simulate a one-day extrapolation due to the limitation of computer

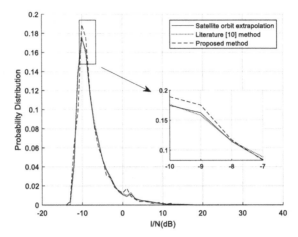

Fig. 9. Comparison of the probability distribution of interference among a 60-day extrapolation, literature [10] method and the proposed method

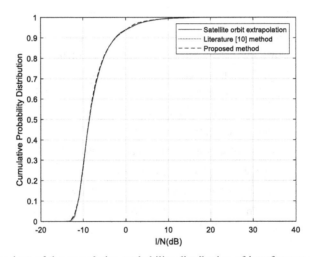

Fig. 10. Comparison of the cumulative probability distribution of interference among a 60-day extrapolation, literature [10] method and the proposed method

Table 3. Simulation duration comparison of the three methods.

Method	Duration
60-day extrapolation	16639 s
Literature [10] method	1253 s
Proposed method	153 s

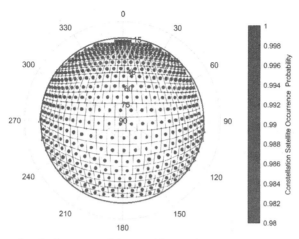

Fig. 11. Division of sub-airspace and its satellite positions and occurrence probability of constellation satellite (elevation/azimuth dimension).

memory, and the final 60-day extrapolation results are obtained by merging multiple extrapolation results. In literature [10], the final probability distribution of interference is calculated by cyclically traversing the interference between the snapshots of interfering and interfered-with constellations. Consequently, the calculation amount depends on the total number of different constellation snapshots and the constellation scale of different snapshots (the snapshots need to be generated by reference satellites). The proposed method does not involve snapshots when calculating mega-constellations. Once the visual airspace is divided into sub-airspaces, the number of sub-airspaces and satellite positions can be determined, thus significantly improving computational efficiency.

Table 4 illustrates the simulation parameters for another interfering constellation systems. It is a multi-orbital altitude constellation, including star constellation and δ

Table 4. Simulation parameters for multiple NGSO interfering-with constellations.

Parameter	Constellation B_1 (star-pattern)	Constellation B_2 (δ constellation)	Constellation B_3 (δ constellation)	Constellation B_4 (star-pattern)
Orbital altitude	1200 km	850 km	1150 km	770 km
Orbital inclination	86°	78°	60°	85°
Number of satellites	7200	5000	3200	6050
Number of orbital planes	60	50	40	55
Parameter	1			
Position of the earth station	[104.066°E, 30.6667°N]			

constellation, with more satellites. In [10], the computational multiplicity rises in the multi-orbital configuration, resulting in a long simulation time. Therefore, this scenario is only compared with the extrapolation method.

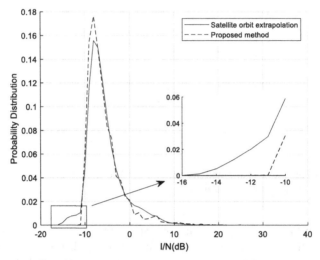

Fig. 12. Comparison of the probability distribution of interference between a 60-day extrapolation and the proposed method (multiple orbital configurations)

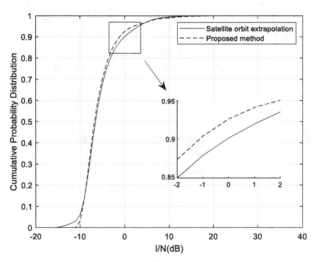

Fig. 13. Comparison of the cumulative probability distribution of interference between a 60-day extrapolation and the proposed method (multiple orbital configurations)

Simulation results are demonstrated in Fig. 12 and Fig. 13, the interference probability distribution curve exists a partial interference value probability is 0 in the region with smaller interference values (magnified region). There is an error of about 0.02 in the cumulative interference probability distribution curve (magnified region). The reason for the error is the same as the above mentioned: because of the uniform distribution of satellites in different sub-airspaces, the situation of smaller interference value cannot be simulated (i.e., sparse distribution of satellites in the co-linear interference direction), but it possess little influence on the cumulative interference probability distribution. Table 5 shows the comparison of the simulation durations of the two methods. It can be observed that although the number of constellation layers and the quantity of satellites are obviously increased, the computation duration of proposed method is not significantly improved, but also keep a high computational efficiency, which is two orders of magnitude higher than the extrapolation method.

In summary, the proposed method approximates the mega-constellation by partitioning the sub-airspaces, sacrificing some accuracy for a significant increase in efficiency. There are some errors, but the errors are small enough. Therefore, the proposed method can be used to quickly calculate the interference involving multiple mega-constellations.

Table 5. The simulation duration of the extrapolation method and the proposed method.

Method	Duration
60-day extrapolation	35775 s
Proposed method	206 s

4 Conclusion

In this paper, we propose a fast method for calculating the interference probability distribution of mega-constellations involving multiple orbital altitudes. According to the occurrence probability of constellation satellite, the visual airspace is divided into several sub-airspaces and combined with the occurrence probability of satellite with different orbital altitudes to calculate aggregate interference from satellites within different sub-airspaces in different altitudes. The interference from dynamically changing satellites is characterized as interference caused by different sub-airspaces, without extrapolating satellite positions or generating constellations by reference satellites, which significantly reduces the computation amount. Simulation results demonstrate that the proposed method can extremely improve the efficiency of calculating the interference probability distribution. This method approximates the mega-constellation by dividing the visual airspace, which inevitable to introduce some errors, but the error is small enough. Hence, this method can be used to fast calculate the interference probability distribution of mega-constellations.

References

1. Attachment Exh 4 Legal Narrative, SAT-MPL-20200526-00053. FCC, Washington, DC (2020)

2. Legal Narrative: SAT-LOA-20190704-00057. FCC, Washington, DC (2020)
3. Attachment Narrative App, SAT-MOD-20200417-00037. FCC, Washington, DC (2020)
4. Mihai, A.: SpaceX GEN2 non-geostationary satellite system, Attachment Narrative App, SAT-LOA-20200526-00055. FCC, Washington, DC (2020)
5. Emilee Speck. https://www.clickorlando.com/news/local/2020/09/01/spacex-starlink-satell ite-constellation-will-reach-700-in-orbit-with-this-weeks-launch/. Accessed 01 Sept 2020
6. Attachment Legal Narrative, SAT-MPL-20200526-00062. FCC, Washington, DC (2020)
7. Gao, X.: Research on dynamic time-varying channel interference mechanism and evaluation technology for space internet constellation system (2019)
8. Fortes, J.M.P., Sampaio-neto, R., Maldonado, J.E.A.: An analytical method for assessing interference in interference environments involving NGSO satellite systems. Int. J. Satell. Commun. **17**(6), 399–419 (1999)
9. Fortes, J.M.P., Sampaio-neto, R., Goicochea, J.M.O.: Fast computation of interference statistics in multiple non-GSO satellite systems environments using the analytical method. IEE Proc. Commun. **151**(1), 44–49 (2006)
10. Lin, Z., Jin, J., Yan, J., Kuang, L.: A method for calculating the probability distribution of interference involving mega-constellations. In: 71th International Astronautical Congress, IAC-2020-B2.1.3, Dubai (2020)
11. Interference analysis to accompany the request for modification of the STEAM-2B non-geostationary satellite system. CRc4422M3 interference analysis with respect to compliance with RoP no. 9.27 STEAM-2B MOD report, p. 29
12. Jin, J., Li, Y., Zhang, C., et al.: Occurrence probability of co-frequency interference and system availability of non-geostationary satellite system in global dynamic scene. J. Tsinghua Univ. (Sci. Technol.) **58**(009), 833–840 (2018)
13. Zhang, C., Jiang, C., Jin, J., et al.: Spectrum sensing and recognition in satellite systems. IEEE Trans. Veh. Technol. **68**, 2502–2516 (2019)
14. Zhang, C., Jin, J., Zhang, H., et al.: Spectral coexistence between LEO and GEO satellites by optimizing direction normal of phased array antennas. China Commun. **15**(006), 18–27 (2018)
15. Zhang, C., Jin, J., Kuang, L., et al.: Blind spot of spectrum awareness techniques in non-geostationary satellite systems. IEEE Trans. Aerosp. Electron. Syst. **54**, 3150–3159 (2018)
16. Li, T., Jin, J., Yan, J., et al.: Resource allocation in NGSO satellite constellation network constrained by interference protection to GEO satellite network. In: 70th International Astronautical Congress, IAC-2019-B2.3.13, Washington D.C., United States (2019)
17. Analytical method to calculate visibility statistics for NGSO satellites as seen from a point on the earth's surface: recommendation ITU-R S.1257. International Telecommunication Union, Geneva (2002)
18. Reference earth-station radiation pattern for use in coordination and interference assessment in the frequency range from 2 to about 30 GHz: recommendation ITU-R S.465-5. International Telecommunication Union, Geneva (2010)
19. Satellite antenna radiation patterns for non-geostationary orbit satellite antennas operating in the fixed-satellite service below 30 GHz: recommendation ITU-R S.1528. International Telecommunication Union, Geneva (2001)
20. Lin, Z., Li, W., Jin, J., Yan, J., Kuang, L.: Research on Satellite Occurrence Probability in Earth Station's Visual Field for Mega-Constellation Systems. In: Yu, Q. (ed.) SINC 2019. CCIS, vol. 1169, pp. 207–220. Springer, Singapore (2020). https://doi.org/10.1007/978-981-15-3442-3_17

Research on Multi-domain Virtual Resource Mapping Method for Aeronautic Swarm Network Supporting Differentiated Services Mechanism

Xiang Wang[(✉)], Jingpeng Ran, Shanghong Zhao, Yiting Nie, and Xinggang Lei

Information and Navigation College, Air Force Engineering University, Xi'an 710077, China

Abstract. For the demand on differentiated services for aeronautic swarm networks, how to achieve the efficient allocation of network resources to meet the requirements of deep-level cooperative information transmission between air combat platforms becomes an urgent problem to be solved in the development of aeronautic swarm network. Aiming at the case that the underlying physical resources of the aviation swarm network are divided into multiple regions, combined with the characteristics of multi-domain resource mapping, the virtual resource mapping strategy based on multi-domain scenario is studied. Firstly, referring to the existing hybrid hierarchical concept, a new mapping architecture of aeronautic swarm network is designed. Under this architecture, public information is considered based on two aspects of physical resource and network topology to realize information sharing among multiple aviation communication Infrastructure Provider (InP-AC). Then considers dividing the multi-domain mapping into three stages: resource matching, inter-domain mapping and intra-domain mapping. Focuses on the inter-domain mapping phase, considers network cost and inter-domain.

Keywords: Aeronautic swarm networks · Differentiated services mechanism · Virtual network mapping method

1 Introduction

Aviation field is the significant component of systematic global battlefield all the time. With the constant development of war thought and specific technology, the evolution of aeronautical operation concept and operation mode is promoted because of the strong uncertainty, high level of dynamicity and high antagonism to environment. Especially with the continuously growing requirement of operational effectiveness in increasingly complicated combat environment, due to the platform load, platform maneuverability, platform stealth, electromagnetic compatibility and other requirements, only relying on a single aviation platform with strong and complete operational capabilities can no longer meet the diversification of task requirements of future air combat [1]. In recent years, researchers have applied the concept of swarm to the field of aviation operations, which inspired by the behavior of biological swarm, and put forward the concept

© Springer Nature Singapore Pte Ltd. 2021
Q. Yu (Ed.): SINC 2020, CCIS 1353, pp. 35–51, 2021.
https://doi.org/10.1007/978-981-16-1967-0_3

of aviation swarm [2–5]. Compared with traditional fleet with platform-centric warfare concept, aeronautic swarm puts more emphasis on network-centric concept. To achieve the goal of complementary advantages of operational capacity between platforms and to break through the operational effectiveness bottleneck of single aerial platform, aeronautic swarm takes advantage of network to establish high-efficient information interaction ability and collects different kinds of combat resources, including computing resources, storage resources, cognitive resources (radar and infrared detection equipment), implement resources (air-to-air missile, electronic jamming pod) [6–8]. Therefore, building an aeronautic swarm network that could meet the demand of future air combat application cluster, polymerizing all kinds of operational resources within and between platforms based on of flexible coupling task to meet requirements of deep-level cooperative information transmission is pivotal for aviation swarm to exert actual combat effectiveness.

Existing network (airborne network) can support a certain degree of operational synergies in the existing air combat mode between platforms in the aspect of transmission reliability, end-to-end delay, transmission rate, but its essence is not based on a background of an operational cluster application. There are many problems existing in the requirements of future aeronautic swarm.

The performance of aeronautic swarm network is tightly coupled with specific network technical standards, battlefield resources and specific missions, which is that the formation process of the strike chain of "discovery, identification, decision, strike and evaluation" is coordinated with the allocation process of various battlefield resources. And according to the characteristics of the task execution object and its interactive business, aeronautic swarm network provides differentiated service for the execution of different tasks by rationally allocating the resources of different battlefields [9]. Taking Fig. 1 as an example, when the video data flow of A in Task 1 and the fire control data flow of F in Task 2 are both routed and forwarded through E, there are obvious differences in the Quality of Service (QoS) requirements for transmitting of fire control data flow and video data flow. If there are no differences in transmission, the QoS requirements of the two data flow may not be satisfied, which will greatly reduce the operational performance. Therefore, it is necessary to provide differentiated network services for the transmission of the two data flow by targeted channel resources allocation. However, the traditional aviation network only realizes the cooperation between platforms, and the interconnection ability between heterogeneous networks is poor. Basically, the traditional aviation network does not have the ability to flexibly provide differentiated network services for different combat missions. Taking aviation data link as an example, existing or under study aviation data link supports fixed air combat mode, such as command and guidance, formation cooperation, etc., which has the relatively fixed capacity demand for data link system [10, 11]. The incompatible communication technology adopted by different data link systems and the inconsistent service capacity of communication transmission cause that the message format and protocol standard designed are not uniform.

Therefore, it is an urgent problem to be solved in the construction and development of aeronautic swarm network that how to realize the flexible coupled task of differentiated service based on the effective allocation of various battlefield resources.

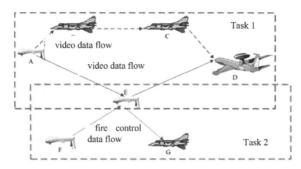

Fig. 1. Aeronautical cluster collaboration example

The key to build a communication network with differentiated service capability is how to choose different parameters such as bandwidth, delay, packet loss rate and so on to ensure multilevel end-to-end QoS. Network Virtualization (NV) is considered as one of the most effective approaches. NV technology can separate the network Infrastructure Provider of Aeronautical Communication (INP-AC) from the Service Provider of Aeronautical Communication (SP-AC) without changing the current basic network architecture. It builds multiple heterogeneous virtual networks based on sharing underlying physical entity resources and different virtual networks are isolated from each other, on which new network technologies can be independently deployed and run to meet diversified business needs [12]. Therefore, this paper puts forward the research on multi-domain virtual resource mapping method of aeronautic swarm network based on the multi-domain cluster air combat scenario, to provide theoretical reference for the construction of aeronautic swarm network supporting differentiated services.

2 Mapping Architecture and Process Analysis

2.1 Mapping Architecture

In single-domain scenarios, a single INP-AC can obtain all the information of physical network, and the decision-making platform can make global optimization based on it, so the mapping efficiency is high. However, in a multi-domain scenario, the underlying network is composed of different physical domains, and the domains for which each physical domain responsible are interconnected by intra-domain boundary nodes through inter-domain links. Network information in different physical domains cannot be shared entirely, only partial information in the domain can be selectively shared. Based on this, this paper draws on the existing hybrid multi-level control architecture [13] to design a new mapping architecture of multi-domain virtual network based on limited information sharing within the domain, as shown in Fig. 2. The separation of data plane and control plane makes SP-AC have no need to directly negotiate with multiple physical domains in the cross-domain mapping, but through the intermediate control layer, and the function of Virtual Network Provider (VNP) is replaced by the control layer.

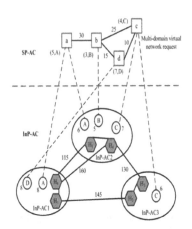

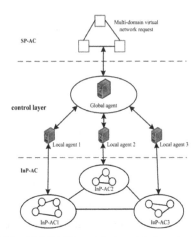

Fig. 2. Resource matching examples

Fig. 3. Hybrid hierarchical multi-domain mapping architecture

In this architecture, SP-AC is responsible for building virtual network requests and providing network services for users. The control layer is divided into global controller and local controller. The global controller is responsible for collecting the inter-domain information of the underlying network and part of the intra-domain information uploaded by the local controller, based on which to construct the global mapping view. Then, it receives the multi-domain virtual network request sent by SP-AC and splits it between domains. After that, it sends the partitioned sub-virtual network request to the local controller, and maps the multi-domain virtual link connecting the sub-virtual network to the inter-domain physical link, and at last, it will confirm whether the mapping is successful. The local controller is responsible for managing the physical domains of each INP-AC and uploads limited shareable physical domain information to the global controller according to the multi-domain mapping target. At the same time, it receives the sub-virtual network requests issued by the global controller, and maps it in the domain. INP-AC is responsible for providing the underlying physical Network resources, which is similar to the data plane under the Software Defined Network (SDN) architecture. It is composed of multiple switch nodes, which are controlled by local controllers and upload all their own information.

2.2 Information Sharing

An efficient and reasonable decision of virtual network mapping requires the global controller to master as much underlying physical network information as possible, although each INP-AC will not disclose all the network information in the physical domain with the purpose of protecting trade secrets [14]. So, this paper draws on some ideas in [15] and considers the underlying physical network information that can be disclosed between each INP-AC based on physical resources and network topology to provide decision support for the multi-domain mapping scheme. According to the resource information disclosure criteria of Amazon EC2 [16], each INP-AC should disclose the resource type and price information of its physical domain. For the underlying network topology, the topological connection between nodes at the boundary of each physical domain and the inter-domain link bandwidth as well as price information will be uploaded to the global controller. In addition, the global controller can obtain the utilization of inter-domain link bandwidth through the local controller, including the consumption and total amount of bandwidth, so that the obtained global view information is more detailed. As shown in Fig. 3, the lower part consists of three physical domains managed by InP-AC, uppercase letters surrounded by circles represent node resource types that InP-AC could provide, and a 'H' with numbers in hexagons is boundary node, the figures next to inter-domain link show the corresponding resource price per unit. What we can learn from Fig. 3 is that the unit price definitions of the same type of physical node resources in different INP-AC physical domains may be inconsistent, and the unit price definitions of link resources between different domains may also be different.

2.3 Mapping Process

Resource Matching Phase: When the global controller receives the multi-domain virtual network request sent from SP-AC, it can find the mapping candidate INP-AC set by executing the matching algorithm [17], according to the requirement description of the underlying physical resources in the request. As shown the upper part of Fig. 3, lowercase letters surrounded by squares represents the virtual nodes, between which is the virtual link, and the symbols in parenthesis represent CPU resources and node type requirements, the numbers beside the link represent bandwidth requirements. For example, candidate InP - AC sets of 'a' is {InP-AC1, InP-AC2} calculated by matching algorithm.

Inter-domain Mapping Phase: When the virtual nodes obtain their candidate INP-AC sets, if any of the underlying physical network INP-AC cannot meet all resource requests, the global controller shall carry out inter-domain mapping according to the obtained candidate INP-AC set information, that is, carry out topological segmentation to the virtual network requests. In the process of segmentation, multi-domain requests are divided into multiple sub-virtual networks corresponding to different candidate INP-AC physical domains, and the virtual links connecting different sub-virtual networks are mapped to inter-domain links. How to segment virtual network requests economically, efficiently and reliably is the core research content of multi-domain mapping algorithm.

Intra-domain Mapping Phase: After the inter-domain mapping is completed, each sub-virtual network request is sent to the local controller, which instructs each INP-AC physical domain to complete the intra-domain virtual network mapping independently. Now, the mapping problem of multi-domain virtual network is transformed into a single-domain mapping problem, and intra-domain mapping can be completed according to the single-domain mapping algorithm.

3 Multi-domain Virtual Network Equilibrium Reliable Mapping Model Description

3.1 Problem Analysis

For each low-level INP-AC, the unit price of physical resources is different, and the unit price of inter-domain links is much higher than that of intra-domain links [18]. In order to reduce mapping spending, virtual nodes should be mapped to INP-AC domains with lower unit price of node resources as far as possible, and virtual links should be mapped to intra-domain links as far as possible. Secondly, in order to ensure the uninterrupted information interaction between sub-virtual network requests, the inter-domain links generally carry a large amount of data traffic, which requires high link reliability, considering the actual situation of the network. In addition, most of existing algorithms aim at mapping spending, the one who has lower unit price is more likely to be the priority, which will make the inter-domain links with the lower unit price have more and more pressure on carrying capacity. And that will result in network congestion, affect the mapping of subsequent requests, and reduce the acceptance rate of requests. Therefore, in order to realize economic and reliable mapping, the reliability and equilibrium of the mapping should be considered, besides the spending.

To reduce the complexity of the problem and optimize the existing multi-domain mapping algorithm, the underlying physical network among InP-AC domain link should be remodeling analysis. Through the study the allocation plan of virtual node-boundary node, the multi-domain virtual network requests can be reasonably divided, which can realize the goal of cost minimization and domain link equilibrium and reliability. What should be noted is that the mapping to the boundary node here is not a bearing virtual node, but only indicates that the virtual node is divided into the INP-AC domain to which the boundary node belongs, and connections are established with other INP-AC domain boundary nodes to realize inter-domain mapping through this boundary node. Since the unit price of inter-domain links is much expensive than intra-domain links, the cost of virtual link can be considered to be 0 if both ends of the virtual links are mapped to the same boundary node. If it is mapped to two different boundary nodes, the link cost is defined as the product of the bandwidth requirement and the link unit price between the two boundary node domains.

3.2 Description of the Underlying Network and Virtual Network Requests

The whole underlying physical network is composed of different INP-AC physical domains connected by inter-domain links between boundary nodes, which can be expressed as $\bigcup_{m=1}^{D} G_s^m (1 \leq m \leq D) \cup L_{s,a}$, D represents the total number of INP-AC physical domains in the underlying physical network, and Ls,a represents the set of inter-domain links. From the global view, the network part within the domain can be approximately ignored, and the underlying network is considered to be composed of boundary nodes and connecting links, which is represented by weighted undirected graph: $G_s = (N_{s,b}, L_{s,a})$, and Ns,b represents a set of boundary nodes. Specifically, the underlying physical network is represented by a weighted undirected graph:

$G_s = \left(N_s, L_s, C_s^N, C_s^L\right)$, and N_s represents the node set in the topology, L_s represents the link set is the attribute set of nodes $n_s(n_s \in N_s)$, including computing source $CPU(ns)$, resource unit price of CPU($UC_N(ns)$). C_s^L is the attribute set of links $l_s(l_s \in N_s)$, including bandwidth resources $b(l_s)$ and unit price of bandwidth resources $UC_L(n_v)$.

Similarly, virtual network requests is represented by $G_v = \left(N_v, L_v, C_v^N, C_v^L\right)$, where N_v and L_v respectively express the set of virtual nodes and links, C_v^N is the resource constrained set of nodes $n_v(n_v \in N_v)$, including virtual nodes requests $CPU(n_v)$, and C_v^L is the attribute set of links $l_v(l_v \in L_v)$, including link bandwidth requirement $b(l_v)$.

3.3 Problem Modeling

When the resource matching stage is completed, the candidate INP-AC set of each virtual node can be obtained, and then a feasible mapping scheme can be constructed according to the candidate set. Based on the ideas in [19], the mapping matrix of virtual nodes corresponding to the physical domain of their boundary nodes is constructed by using the correspondence between virtual nodes and boundary nodes. The matrix is represented as $Q_{m \times n}$, m is the number of virtual nodes in the request, and n is the number of boundary nodes in the underlying network. For the candidate set constructed by each virtual node in Fig. 2, a feasible mapping scheme can be expressed in Table 1:

Table 1. A feasible mapping scheme for multi-domain virtual requests

	H_1	H_2	H_3	H_4	H_5	H_6
a	1	−1	−1	0	0	0
b	−1	−1	−1	1	0	−1
c	−1	1	0	0	0	−1
d	0	−1	−1	−1	−1	1

In this table, the value of Q_{ij} represents the mapping relationship between the virtual node and the physical domain of its boundary node.

$$
Q_{vj} = \begin{cases}
1, & \text{border node } j \text{ is } in \text{the mapping range} \\
& \text{of virtual node } v \text{ and } v \text{ maps to } j \\
0, & \text{border node } j \text{ is } in \text{the mapping range} \\
& \text{of virtual node } v \text{ but } v \text{ fails mapping to } j \\
-1, & \text{border node } j \text{ is without } in \text{the} \\
& \text{mapping range of virtual node } v
\end{cases}
\tag{1}
$$

Since each virtual node can map to only one physical domain, that is, to a unique boundary node, so each row has one and only one element that takes 1. According to the definition of Q_{ij}, there is a unique mapping scheme for a specific Q_{ij} matrix corresponding to it, as for a feasible mapping scheme, the number of -1 and 0 in the matrix has been determined. The successful mapping of each node changes the element from 0 to 1 corresponding to the unique Q_{ij} matrix, which proves the rationality of the model construction.

While considering the mapping spending of multi-domain virtual network, the reliability and equilibrium of inter-domain links are also considered. In order to minimize the spending under the premise of maintaining the reliability and equilibrium of network, an integral linear programming model is established, which is defined as follows:

Node Spending. The spending C_j^i mapping from virtual node n_i to boundary node n_j can be represented by follows:

$$C_j^i = \sum_i \sum_j X_N(i,j) \cdot CPU(i) \cdot UN_j(i) \tag{2}$$

Link Spending. In this paper, inter-domain link spending is used to estimate the entire link mapping spending, and the spending of virtual link l_{uh} mapping to inter-domain link l_{ij} can be expressed as follows:

$$C_{ij}^{uh} = \sum_{ij} \sum_{uh} M_L(uh, ij) \cdot b(l_{uh}) \cdot UC_L(l_{ij}) \tag{3}$$

Link Unbalance Degree. Due to the frequent occurrence of link failure in aviation network environment, it is necessary to pay more attention to ensure the reliability and stability of network links. Firstly, the link load is defined according to the utilization of inter-domain link bandwidth. Load $(l_{s,a})$ is the ratio of the sum of all virtual link bandwidth demand resources mapped to the inter-domain link in the request to the original bandwidth resources of the link. Secondly, by referring to the idea of statistical analysis, the inter-domain link reliability is defined as the ratio of the total number of interruptions in the whole underlying network to the number of interruptions in any physical link $l_{s,a}$, which is expressed by θ. The link imbalance degree is defined in this paper as shown in Eq. (6):

$$load(l_{s,a}) = \sum_{\forall l_v \in L_v \rightarrow l_{s,a}} b(l_v)/b(l_{s,a}) \tag{4}$$

$$\theta = \sum_{l_{s,a} \in L_{s,a}} f(l_{s,a})/f(l_{s,a}) \tag{5}$$

$$D_e = \frac{\max\limits_{l_{s,a} \in L_{s,a}} load(l_{s,a})}{\theta \cdot \sum_{l_{s,a} \in L_{s,a}} load(l_{s,a})/|L_{s,a}|} \tag{6}$$

In conclusion, the aim of considering network spending and reliable equilibrium while defining the objective function is to obtain a multi-domain mapping scheme that minimizes the spending and link imbalance degree. The specific expression is shown in Eq. (7):

$$\text{Min} f = \frac{\alpha\left(C_j^i + C_{ij}^{uh}\right)}{2 \times 10^{\mu_1}} + \frac{\beta D_e}{2 \times 10^{\mu_2}} \tag{7}$$

α, β is the adjustment coefficient aiming to balance mapping overhead and network imbalance degree; μ_1, μ_2 is the smallest integer where the adjusted spending and imbalance value are located at $(0,1)$, and its purpose is to normalize the function value. At the initial stage of mapping, α is increased to reduce the mapping cost for the load pressure of network is at a low level; at the later stage, β is increased to make the underlying network balanced and reliable as far as possible for the load and the inter-domain link pressure is gradually increasing.

The CPU resources and link bandwidth resources of each INP-AC physical domain node is relatively rich, which can better meet the resource requirements of the virtual network. Therefore, the bandwidth resources constraints of link and CPU of intra-domain node are ignored in the process of inter-domain modeling, and the mapping constraints, inter-domain bandwidth constraints and connectivity constraints are mainly considered, which are specifically expressed as follows:

$$\forall n_i \in N_v, \sum_{n_j \in N_{s,b}} X_N(i,j) = 1 \tag{8}$$

$$\begin{cases} \forall l_{ij} \in L_{s,a}, \sum_{l_{uh} \in L_v} M_L(uh, ij) \cdot b(l_{uh}) \leq b(l_{ij}) \\ b(l_{uh}) \leq b\left(l_{ij}\right) - \sum_{\forall l_v \in L_v \to l_{ij}} b(l_v) \end{cases} \tag{9}$$

$$\forall n_i \in N_{s,b}, \forall l_{ij} \in L_{s,a}, \sum_{l_{ij} \in L_{s,a}} M_L(uh, ji) - \sum_{l_{ij} \in L_{s,a}} M_L(uh, ij) = \begin{cases} 1, & X_N(u,j) = 1 \& \\ & X_N(h,j) = 0 \\ -1, & X_N(h,j) = 1 \& \\ & X_N(u,j) = 0 \\ 0, & else \end{cases} \tag{10}$$

Equation (8) means that any virtual node can only be mapped to one boundary node; Eq. (9) indicates that the bandwidth demand of any virtual link, of which will be mapped to the inter-domain link, is less than the available bandwidth of the inter-domain link; Eq. (10) indicates that if the two endpoints u and h of link l_{uh} are mapped to the same boundary node n_j, then both the inflow and outflow flows are 0, which means that l_{uh} will carry out single-domain mapping in the physical domain that the boundary node n_j.

4 Optimization Strategy Design of Inter-Domain Mapping Based on Ant Colony Algorithm

In order to obtain the approximate optimal solution of multi-domain virtual network partition, the ant colony algorithm is used to optimize the virtual node-boundary node selection in the virtual network request, and the virtual network is divided into several sub-virtual networks. In view of the drawbacks of the traditional ant colony algorithm [20], such as "precocity", slow convergence, and fall into local extreme value easily, this paper introduces adaptive state transition and pheromone update strategy based on elite selection to improve the traditional algorithm.

4.1 The Problem Coding

The algorithm uses the state of each ant to represent a solution to the problem, a possible partition scheme, which is represented by $X = [x_{i1}, x_{i2}, \ldots, x_{id}]$, the solution containing corresponds to virtual node requests which is d, x_{ij} is the relationship of virtual nodes and boundary nodes mapping, its value is a virtual node j maps to boundary nodes x_{ij}, the boundary nodes selection of each virtual node constitute classification scheme, and X is the calculation of the objective function $f(X)$, to evaluate the advantages and disadvantages of the scheme.

4.2 Probability of Adaptive State Transition

In the construction process of the feasible solution, it is assumed that the number of iterations is k, and each ant selects the boundary node according to the information of k-1 iteration. There are two ways for each virtual node of the ant to select the boundary node:

$$x_{ij} = \begin{cases} \arg \max_{bq \in N_{s,b}^c} \left(\left[\tau_{ij,bq}(k)\right]^{\chi^1} \cdot \left[\eta_{ij,bq}(k)\right]^{\chi^2} \right), & e \leq p \\ x_{wp}, & e > p \end{cases} \tag{11}$$

$\tau_{ij,bq}(k)$ is the information concentration of generation k, $\eta_{ij,bq}(k)$ is heuristic information, χ^1, χ^2 is the adjusting parameters of each weight, $N_{s,b}^c$ is the optional boundary node candidate set of virtual node j, p is the adaptive selection ratio, and e is the random generation number.

When the number of random generation e does not exceed p, the boundary node with the largest pheromone with x_{ij} is selected as the pre-mapped boundary node. Otherwise, roulette is applied to select according to the probability of state transition, as follows:

$$p_{ij,bq} = \frac{\left[\tau_{ij,bq}(k)\right]^{\chi^1} \cdot \left[\eta_{ij,bq}(k)\right]^{\chi^2}}{\sum_{bq \in N_{s,b}^c} \left(\left[\tau_{ij,bq}(k)\right]^{\chi^1} \cdot \left[\eta_{ij,bq}(k)\right]^{\chi^2} \right)}, \quad bq \in N_{s,b}^c \tag{12}$$

The heuristic information $\eta_{ij,bq}(k)$ needs to be redefined according to the actual situation of link l between x_{ij} and x_{bq}, and the boundary nodes with low link load and high reliability are selected for access, as shown in Eq. (13). ξ is weight regulation coefficient.

$$\eta_{ij,bq}(k) = \frac{\theta(l)^\xi}{load(l)} \tag{13}$$

As for the determination of the adaptive selection proportion p, the current iteration number is calculated in an adaptive way, and its value changes dynamically with the change of the value of IT, as shown in Eq. (14).

$$p = p_{max} - \frac{It_{cur}}{It_{max}} \cdot (p_{max} - p_{min}) \tag{14}$$

It_{cur} is the number of current iterations, It_{max} is maximum number of iterations, p_{max} and p_{min} is the upper and lower limits of the adaptively select ratio. At the beginning of iteration, the value of q is large and the convergence speed is faster. In the late iteration, the value of q is small, and the diversity of the population increases.

4.3 Pheromone Update Strategy Based on Elite Selection

In the process of ant iterative optimization, a pheromone updating strategy based on elite selection is proposed in order to make the optimal path rapidly converge within the nearest range of the better solution. For each searched path, it sorts the path in reverse order according to the value of the objective function from small to large, and different pheromone updating methods are adopted for the paths with different serial numbers.

If the sequence number of the path is in the first 1/3 of this cycle, it is regarded as an elite solution, and its pheromone is further strengthened. If the sequence number of the path is between 1/3 and 2/3 of the cycle, the global pheromone will be updated. If the sequence number of the path is in the last 1/3 of the cycle, its pheromone is additional weakened. The pheromone update formula combined with the elite strategy is as follows:

$$\tau_{ij,bq}(K + 1) = (1 - \rho) \cdot \tau_{ij,bq}(k) + \Delta\tau_{ij,bq}(k) \tag{15}$$

$$\Delta\tau_{ij,bq}(K) = \begin{cases} \frac{Q}{f(r)/f_{max}}, & p_{rank} < \frac{NI}{3} \\ 0, & \frac{NI}{3} < p_{rank} < \frac{2NI}{3} \\ -\frac{Q}{f(r)/f_{max}}, & otherwise \end{cases} \tag{16}$$

$\rho \in [0, 1]$ is pheromone volatility coefficient, $\Delta\tau_{ij,bq}(K)$ is additional pheromone increments of link $p(ij, bq)$, NI is the ant scale, Q is pheromone enhancer, τ_{max} and τ_{min} is upper and lower limits of pheromones which are set on the solution path in each generation of elite solution.

4.4 Design of Interdomain Equalization and Reliable Mapping Algorithm

The main idea of algorithm design is: first, choosing a feasible virtual network classification scheme as the ant optimization of the initial solution based on each candidate InP – AC set of each node obtained by resources matching phase.

Then the goal function, combined with the constraint conditions of ant optimization process, finds the right boundary for each virtual node and obtains multi-domain virtual network dividing approximate optimal solution through continuous iteration. A specific algorithm is described in Table 2.

Table 2. Virtual network partition algorithm based on optimized ant colony algorithm

---

OAC-MVNE algorithm

---

input : Multi-domain virtual network requests $G_v=(N_v,L_v,C_v^N,C_v^L)$, physical network $G_s=(N_s,L_s,C_s^N,C_s^L)=(N_{s,b},L_{s,a})$

output: Results of sub-virtual network partition

1:      Set relevant parameters

2:      Initialize ant colony, pheromone matrix // The storage matrix of pheromone is a two-dimensional matrix which is determined by the number of virtual nodes and the number of boundary nodes

3:      Calculate transfer probability according to heuristic information and pheromone

4:      **for** n=1 to Nl **do**

         According to the resource matching algorithm, select a feasible partition scheme as the initial solution X_i

5:          Calculate the objective function value $f(X_i)$

6:          $X_{best}=X_i$

7:      **end for**

8:      **for** m=0 to It **do**

9:          **for** n=1 to Nl **do**

10:            Updated the ant colony according to the pheromone, and generate the virtual node-boundary node partition scheme and update X_i

11:            **if** $f(X_{best}) > f(X_i)$

12:                $X_{best}= X_i$

13:            **end if**

14:        **end for**

15:        Update heuristic information, transition probability, and pheromone matrix

16:        **if** $n> Nl$

17:            **Break**;

18:        **end if**

19:    **end for**

20:    **if** ant X_{best} satisfies all node mapping constraints and can use K-shortest path method to find all paths that conform to inter-domain link bandwidth constraints and connectivity constraints

21:        Record node mapping and link mapping results

22:        According to the result of mapping, the virtual network is divided to form the sub-virtual network set G_v^m

23:        update the various resources in G_s

24:        **return** Divide successfully

25:    **else**

26:        **return** Divide unsuccessfully

---

5   Simulation and Result Analysis

The underlying physical network and virtual network requests are randomly generated by the GT-ITM topology generator [21] in the simulation experiment network topology. The underlying physical network contains 5 autonomous domains, and each domain contains an average of 20 nodes and 80 links. A total of 100 physical nodes and 400 physical links are distributed randomly and uniformly. Both node CPU resources and intra-domain link bandwidth resources obey the uniform distribution of [50,100]. There are two boundary nodes in each domain, which are fully connected with each other with a connection probability of 0.5. The bandwidth resources of boundary links obey the uniform distribution of [200,300], the unit price of CPU resources of physical nodes obey the uniform distribution of [1, 10], and the unit price of bandwidth resources of inter-domain links obey the uniform distribution of [100, 200]. Virtual network request arrival simulatively generates according to Poisson distribution mode. A total of 2,000 virtual network mapping requests are generated in 50000-time units, and the arrival requests is 4 in the average number of 100-time units, and the life cycle of each virtual network request follows an exponential distribution with an average of 200-time units. The number of virtual nodes follows the uniform distribution of [2, 12], and the probability of connection between nodes is also 0.5. The demand for CPU resources and link bandwidth resources of virtual nodes follows the uniform distribution of [1, 15] and [5,30], respectively. Inter-domain link interruptions follow a Poisson process with an intensity of 4. The ant parameters in the experiment are set as follows: population size NI is 50, the number of iterations is 1000, χ_1 and χ_2 is 0.5 and 1.5, ξ is 3, the volatility coefficient ρ is 0.1, pheromone enhancer Q is 0.1, $\tau \in [0.02, 1]$, $\tau_0 = 0.1$, $p \in [0.1, 0.9]$, α and β both are 0.5 in Eq. (7).

The algorithm uses the generated same network topology to evaluate the optimization algorithm from three performance scales, including average partition time, request acceptance rate and mapping spending, and compares with the algorithms LID-MVNE [22] and Polyvine [23]. In order to investigate the performance of OAC-MVNE algorithm and facilitate the comparison, the domain mapping algorithms of several algorithms all adopt greedy strategy.

Figure 4 shows the variation of the objective function value with the number of iterations after adopting OAC-MVNE algorithm. As seen in the figure, the optimization objective gradually converges, and better optimization results can be obtained in the later iteration with the continuous optimization process. When the number of iterations is less than 400 generations, the algorithm goes through several rapid optimization processes, and the target value changes obviously. After 400 generations, the target value has experienced multiple progressive optimizations, and the target value gradually tends to be stable, which verifies the effectiveness of the algorithm proposed in this paper. The reason is that the OAC-MVNE algorithm adopts the adaptive state transfer strategy in the boundary node selection, and the value of the adaptive selection ratio changes dynamically with the change of the number of iterations, so that the algorithm can adaptively improve the convergence speed, increase the diversity of the population, and has a better global optimization ability.

Figure 5 compares the variation of average partition time with the number of virtual nodes under different algorithms. Average partition time refers to the required time for the algorithm to partition the same virtual network request, which reflects the efficiency of the algorithm. The figure shows that the LID - MVNE algorithm and PolyViNE algorithm is about exponential growth trend. The average divided time of the OAC - MVNE algorithm is slightly bigger than the LID - MVNE algorithm while the virtual node number is minor (about 6). But when the number of virtual nodes increased (greater

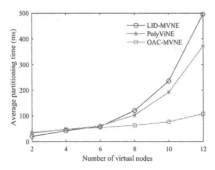   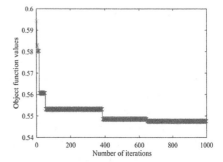

Fig. 4. Average partitioning time   **Fig. 5.** Convergence process

than 6), divided time is about a linear growth trend, when the number is more than eight, OAC - MVNE algorithm possesses the highest execution efficiency obviously.

This is because LID-MVNE algorithm adopts the exact formula strategy, which needs to select the optimal scheme according to the finite open boundary information of the whole situation of partition. The distributed protocol introduced by PolyViNE algorithm coordinates and participates in the mapping process of each INP-AC domain, and ensures the profit maximization by bidding, at the same time the algorithm complexity is high.

OAC-MVNE algorithm simplifies the complex process of multi-domain mapping, and obtains the approximate optimal solution of multi-domain virtual network request partition through heuristic ant colony optimization algorithm on the basis of simple network modeling, so its efficiency is the highest. At the same time, it can be concluded that LID-MVNE algorithm is suitable for small virtual network requests, and OAC-MVNE algorithm is suitable for large virtual network requests.

Figure 6 compares the changes over time of network mapping spending under different algorithms. Before 20,000-time units, the mapping spending of the three algorithms gradually increases and then tends to be stable. The mapping spending of OAC-MVNE algorithm maintains around 18016, which is 200% lower than that of PolyViNE algorithm (about 54120), and LID-MVNE algorithm maintains near 15152 with the lowest overhead. This is because the PolyViNE algorithm is considering to coordinate the node that cannot complete intra-domain mapping to the adjacent InP – AC physical domain mapping, causing the ignores of inter-domain link. And mapping spending is closely related to the inter-domain link costs, which lead to the big spending of virtual network request. The LID - MVNE algorithm and OAC - MVNE algorithms are aimed at mapping spending, the exact type of strategy that the former taken is more efficient at the expense of dividing time, the later makes mapping spending increased slightly according to calculate the approximate solution. In addition, it shows that the mapping spending of the PolyViNE algorithm and LID-MVNE algorithm tends to decrease at the end of the iteration, while the OAC-MVNE algorithm is relatively stable. The main reason is that the network load pressure increases with the continuous progress of mapping, and the PolyViNE algorithm and LID-MVNE algorithm do not consider the balance and reliability of inter-domain links, so that the subsequent virtual network request mapping fails, and the spending is also reduced.

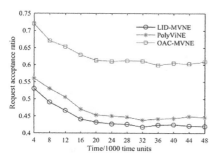

Fig. 6. Request acceptance rate

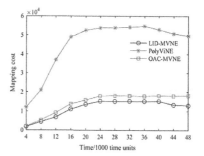

Fig. 7. mapping spending

Figure 7 compares the virtual network request acceptance rate over time under different algorithms. As shown the figure, the request acceptance rate of OAC-MVNE algorithm is the highest, which maintains around 0.60, which is 42.86% and 36.37% higher than that of LID-MVNE algorithm (about 0.42) and PolyViNE algorithm (about 0.44) respectively. This is because the OAC - MVNE algorithm not only consider mapping spending, but also consider the inter-domain link load balancing, reliability and stability. Solving the problems of the virtual network segmentation by ant colony optimization algorithm makes the arrival of the subsequent request virtual network has a more balanced between the domain network, reduces the mapping spending, and improves the acceptance success rate.

In order to analyze the stability of the algorithm, the number of autonomous domains in the underlying physical topology is set to 3 and 7 respectively, and the number of nodes and links in the domain and the corresponding attribute values are set to be consistent, with which in the case of 5 autonomous domains, to verify the stability of the algorithm in different autonomous domains. As seen in Fig. 8(a), in the case of different number of autonomous domains of the underlying topology, the execution efficiency of the OAC-MVNE algorithm has little difference with the increase of the number of virtual request nodes. In general, the average partition time of the algorithm increases slightly with the increase of the number of autonomous domains of the underlying topology. Figure 8(b) compares the mapping spending when the OAC-MVNE algorithm tends to be stable in the case of different number of autonomous domains. It can be seen from the figure that the mapping spending calculated in the three network topologies does not change much, and in general, the mapping spending of the algorithm increases slightly with the increase of the number of autonomous domains. In conclusion, OAC-MVNE algorithm has good stability in different scale network environment.

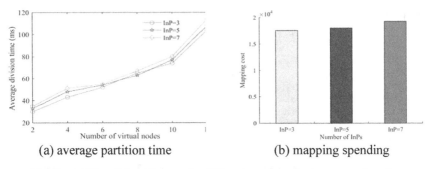

(a) average partition time   (b) mapping spending

Fig. 8. stability of algorithms under different number of autonomous domains

6 Conclusions

The reliable virtual resource mapping method for aeronautic swarm networks based on multiple autonomous domains is proposed in this paper. Based on the existing network virtualization architecture, a new multi-domain virtual network mapping architecture is designed, and the information sharing mechanism among multiple INP-ACs is studied. Then aeronautic swarm network is divided into multi-domain mapping resources matching, inter-domain mapping and intra-domain mapping. Focusing on a single link frequent interruptions case on inter-domain mapping phase, balanced and reliable mapping strategy in the multiple domain OAC – MVNE is designed that based on optimized ant colony algorithm. Optimized ant colony algorithm is used to search for the optimal solution, and the approximate optimal partition scheme is obtained through continuous iteration, and the effectiveness and stability of the algorithm are analyzed.

References

1. Xiaolong, L.: UAV swarm. Northwestern Polytechnic University Press, Xi'an (2018)
2. Niu, Y.F., Xiao, X.J., Ke, G.Y.: Operation concept and key techniques of unmanned aerial vehicle swarms. Natl. Def. Sci. Technol. **5**, 37–43 (2013)
3. Da-jun, H.: Research on Networked Swarm Operation. National Defense University Press, Beijing (2013)
4. Haibin, D., Pei, L.: Autonomous control for unmanned aerial vehicle swarms based on biological collective behaviors. Sci. Technol. Rev. **35**(7), 17–25 (2017)
5. XiaoLong, L., LyuLong, H., JiaQiang, Z., Peng, B.: Configuration control and evolutionary mechanism of aircraft swarm. SCIENTIA SINICA Technologica **49**(03), 39–49 (2019)
6. Shanghong, Z., Kefan, C., Na, L., Xiang, W., Jing, Z.: Software defined airborne tactical network for aeronautic swarm. J. Commun. **38**(8), 140–155 (2017)
7. Huaxin, Q., Haibin, D.: From collective flight in bird flocks to unmanned aerial vehicle autonomous swarm formation. J. Univ. Sci. Technol. **39**(3), 317–322 (2017)
8. Kefan, C., Na, L., Shanghong, Z., Xiang, W., Jing, Z.: A scheme for improving the communications efficiency between the control plane and data plane of the SDN-enabled airborne tactical network. IEEE Access. **99**, 1–16 (2018)
9. Li-ping, H., Xiao-long, L., Jia-qiang, Z., Ping-ni, L.: Research on aircraft swarms system construction mechanism. Fire Control Comm. Control **42**(11), 142–145 (2017)

10. Na, L., Kefan, C.: Aeronautic swarm cloud network oriented MAC protocol for aviation data link. Syst. Eng. Electron. **38**(5), 1164–1175 (2016)
11. Na, L.: Data Link Theory and System, 2nd edn. Publishing House of Electronics Industry, Beijing (2018)
12. Chowdhury, N.M.M.K., Boutaba, R.: A survey of network virtualization. Comput. Netw. **54**(5), 862–876 (2010)
13. Bhole, P., Puri, D.: Distributed hierarchical control plane of software defined networking. In: International Conference on Computational Intelligence and Communication Networks, pp. 516–522 (2015)
14. Dietrich, D., Rizk, A., Papadimitriou, P.: AutoEmbed: automated multi-provider virtual network embedding. In: ACM SIGCOMM Computer Communication Review, pp. 465–466 (2013)
15. Liao, J., Feng, M., Li, T.: Topology-aware virtual network embedding using multiple characteristics. KSII Trans. Internet Inf. Syst. **8**(1), 145–164 (2014)
16. Amazon EC. Instance Types (2012). http://aws.amazon.com/ec2/instance-types
17. Medhioub, H., Houidi, I., Louati, W.: Design, implementation and evaluation of virtual resource description and clustering framework. In: IEEE International Conference on Advanced Information Networking and Applications, pp. 83–89 (2011)
18. Dietrich, D., Rizk, A., Papadimitriou, P.: Multi-domain virtual network embedding with limited information disclosure. In: 2013 IFIP Networking Conference (2013)
19. Feng, L.: Node-selfishness and resources allocation analysis in wireless networks, Xidian University (2017)
20. Hua, X., Zheng, J., Hu, W.: Ant colony optimization algorithm for computing resource allocation based on cloud computing environment. J. East China Norm. Univ. (Nat. Sci.) **1**(1), 127–134 (2010)
21. Liu, J., Ding, F., Zhangx, D.: A hierarchical failure detector based on architecture in VANETs. IEEE Access **99**, 1–1 (2019)
22. Dietrich, D., Rizk, A., Papadimitriou, P.: Multi-domain virtual network embedding with limited information disclosure. In: IFIP Networking Conference (2013)
23. Samuel, F., Chowdhury, M., Boutaba, R.: Polyvine: policy-based virtual network embedding across multiple domains. J. Internet Serv. Appl. **4**(1), 1–23 (2013)

Joint Scheduling of Spectrum and Storage Resources for Content Distribution in Space-Ground Integrated Information Network

Yang Cao[✉], Disheng Wu, Yulong Chen, and Wei Wang

School of Electronics Information and Communications, Huazhong University of Science and Technology, Wuhan, Hubei 430074, China
ycao@hust.edu.cn

Abstract. Aiming at the problem of spectrum resource shortage caused by the rapid increase of multimedia data traffic in terrestrial network, a joint scheduling of storage resources and spectrum resources (JSSS) algorithm in an edge computing enabled space-ground integrated information network (SGIN) is proposed to cache and distribute contents, which decouples communication resources from storage resources by means of phased scheduling. In the Multicast Transmission (MT) stage, we match users and base stations (BSs) or satellites based on a heuristic genetic algorithm. In the cache placement (CP) stage, we place contents adaptively based on a multi-agent deep reinforcement learning (MADRL)-based algorithm. Extensive simulations show that compared with traditional methods, the proposed joint scheduling algorithm can effectively improve the cache hit ratio and reduce the bandwidth consumption.

Keywords: Content distribution · SGIN · JSSS · Cache placement

1 Introduction

Moving into the 5G Era, the global mobile traffic dominated by multimedia data such as video has increased rapidly [1]. In the meantime, the further development of mobile communication is faced with huge challenges. According to the Internet White Paper released by Cisco in February 2019, the global mobile data traffic in 2022 will be 7 times that of 2017, 82% of which will be video traffic. Particularly, the live video traffic will increase 15 times from 2017 to 2022 [2]. The rapid growth of mobile traffic can be caused by the following two aspects: on the one hand, the increasing popularity of users' mobile devices such as smart phones, laptops and the upgrading of hardware; On the other hand, the service demands of users are gradually diversified, such as Augmented Reality (AR) [3], whose requirements for Quality of Experience (QoE) tend to be strict [4, 5]. To solve the above problems, Mobile Edge Computing (MEC) technology is attractive to

This work was supported in part by the National Natural Science Foundation of China (NSFC) with Grants 61729101, 61720106001 and 91738202.

© Springer Nature Singapore Pte Ltd. 2021
Q. Yu (Ed.): SINC 2020, CCIS 1353, pp. 52–69, 2021.
https://doi.org/10.1007/978-981-16-1967-0_4

use in-network caching. MEC can effectively utilize the idle resources at the edge of the network and transfer the load of the backbone network to the edge network, thus improving the network performance and further guaranteeing the users' experience [6].

There are some challenges to cache contents such as video at the base station (BS). First, the MEC server load and user context information (such as channel state, workload and user request patterns, etc.) fluctuate over time and space, making it difficult to design high-performance cache placement strategies. Secondly, in the 5G Era, new services emerge in an endless stream, especially the data scale of multimedia grows explosively. For the demand of surging data transmission at peak hours, it may still be difficult to simply cache data at edge. Therefore, a single MEC technology is not enough to cope with the challenge of spectrum resource shortage in the 5G Era, unable to effectively guarantee the user's experience. With the help of high-performance MEC, the Quality of Service (QoS) of terrestrial network based on base station can be improved traditionally, while the top-down propagation characteristics of satellite networks can provide users with a wider range of coverage, especially in the form of multicast and broadcast. Considering the original advantages of satellite networks and the shortage of spectrum resources of terrestrial network, it is a promising research direction to integrate the two kind of networks to form Space-Ground Integrated Information Networks (SGIN) [7]. However, the existing literature works usually consider the resource optimization in a single domain (i.e., the resource optimization strategy in the storage domain or the spectrum domain), and the coupling of the storage domain and the spectrum domain was less considered.

Motivated by this, in this paper, we study the intelligent scheduling of network resources (i.e., storage and spectrum resources) for the edge computing enabled SGIN and propose a joint scheduling of storage resources and spectrum resources (JSSS) algorithm to maximize the system utility. We solve the optimization problem with a phased strategy: in the cache placement (CP) stage, we utilize a multi-agent deep reinforcement learning (MADRL)-based algorithm to place contents adaptively without knowing the future data popularity or user request patterns, in the multicast transmission (MT) stage, we match users and BSs or satellites based on a genetic algorithm. The contributions of this paper are as follows:

(1) We are the first to consider the coupling of storage resource and the spectrum resource in an edge computing enabled SGIN, aiming at maximize the system utility;
(2) We decouple communication resources from storage resources with a phased strategy and divide the complete scheduling process into two stages, where the storage resources are scheduled with a MADRL-based algorithm in the CP stage and the spectrum resources are scheduled with a heuristic genetic algorithm in the MT stage.
(3) The proposed scheme is simulated and compared from multiple aspects, and extensive simulations show that the proposed scheme can achieve super performance.

2 Related Work

Compared with the high earth orbit (HEO) satellite, the low earth orbit (LEO) satellite has the characteristics of low cost, low delay and large capacity, which is not only the focus of domestic and foreign research institutions [8, 9], but also the most important part of the construction of the access network in SGIN. There are many scholars studying resource scheduling technology in SGIN, which is mainly focused on communication resource scheduling and storage resource scheduling.

In term of communications resource scheduling in SGIN, cell edge users become bottleneck of communication caused by channel fluctuations. Jiang et al. proposed that the satellite and base station share spectrum multicast to assist base station to distribute the contents in order to improve the system performance [10]. In [11], Peng et al. considered two user access modes, unmanned aerial vehicle (UAV) and LEO satellite, and used the evolutionary game theory to solve the problem of user access mode selection in disaster areas in the SGIN. It can be seen that the use of the SGIN can realize the information service with global coverage, provide all kinds of mobile and broadband services, solve the communication transmission and break down the "information island". In term of storage resource scheduling in SGIN, Wu et al. researched the combination of satellites and ground stations cache placement strategy [12]. They designed satellites and stations as two layers cache architecture of content distribution for the user, and optimized the traffic between satellites and users or Internet. Liu et al. considered LEO as cache nodes to provide content distribution for users [13]. After determining data such as file popularity, satellite cache capacity and network topology structure, the average access delay of users was optimized. This paper proved this problem is a Non-deterministic Polynomial Complete (NPC) problem, which can be solved by the proposed exchange stable matching algorithm. Gui et al. considered the cache placement strategy based on social relations in SGIN, and selected the corresponding device-to-device (D2D) cache of each base station to assist users by calculating the "social distance", and deployed the cache in these user terminals to improve the user experience [14]. The process of solution was divided into two stages. First, an appropriate D2D was selected to assist the user, and then a heuristic algorithm was used to place the cache. However, existing works rarely consider the joint optimization of terrestrial network and satellite network resources with the scheduling process of storage resources and communication resources in SGIN, so they cannot accurately describe the characteristics of resource coupling in the SGIN resource scheduling.

In terms of the edge cache placement, the particularity of Internet of Things (IoT) data lies in its transient nature, i.e., the data is only valid during a certain time and will be discarded when it is out of date. At present, most literature works considered caching non-transient data, and assumed that the patterns of user requests are known, which cannot match the essential characteristics of transient data in the IoT. Aiming at the shortcomings of the existing works, Zhu et al. proposed an edge caching strategy for transient data in the IoT based on the deep reinforcement learning [15]. This strategy took into account the transient characteristics of the IoT data in the absence of prior knowledge such as the popularity of the IoT data and the patterns of user requests, which used deep reinforcement learning technology independently to learn historical data and environmental information and intelligently solve the cache placement problem. In addition, an intelligent caching strategy based on deep reinforcement learning was proposed in MEC [16]. The results of simulations based on real data indicated the effectiveness and great potential of the cache strategy based on the deep reinforcement learning in MEC. However, most of the existing works considered only a single edge node, which cannot meet the requirements of the joint cache of multiple edge nodes in reality. At present, the research work of the MEC cache strategy based on MADRL is in the initial stage [17]. In this paper, we propose a multi-agent deep reinforcement learning (MADRL)-based algorithm to place contents adaptively.

In this paper, considering the intelligent joint scheduling process of storage resources and communication resources, JSSS algorithm is designed in an edge computing enabled SGIN. In the CP stage, the base station nodes and the satellite nodes adaptively cache data based on MADRL according to the historical content popularity and other situational information. In the cache placement (CP) stage, we utilize a multi-agent deep reinforcement learning (MADRL)-based algorithm to place contents adaptively without knowing the future data popularity or user request patterns. In the multicast transmission (MT) stage, we match users and BSs or satellites based on a genetic algorithm. The CP stage and the MT stage are cyclically executed to optimize system utility in the long term.

3 System Model and Problem Formulation

3.1 System Model

See Fig. 1.

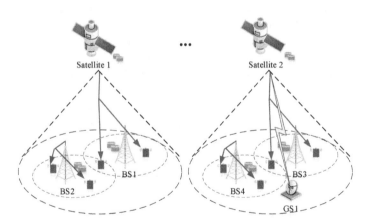

Fig. 1. The system model

As showed in Fig. 2, there are four different types of nodes in SGIN: LEO satellite nodes, base station nodes, Ground Station (GS) nodes and user nodes. We denote the numbers of base station nodes, LEO satellite nodes, ground stations and users as M, R, J and N, respectively. The set of base station nodes, satellite nodes, ground stations nodes and user nodes can be recorded as $B = \{B_m | m = 1, 2, \ldots, M\}$, $S = \{S_r | r = 1, 2, \ldots, R\}$, $G = \{G_j | j = 1, 2, \ldots, J\}$ and $U = \{U_i | i = 1, 2, \ldots, N\}$, respectively. In the following paper, satellite nodes and base station nodes are referred to edge nodes.

Base Station Communication Model: For base station mode, the Signal to Noise Ratio (SNR) of user i communicating with base station node m can be written as:

$$B_{im}^{SNR} = \frac{S_i^{B_m}}{N_i^{B_m}} = \frac{P_b \cdot H_{im}}{N_0}, \tag{1}$$

where P_b is the transmitting power of the base station node, H_{im} is the channel gain and N_0 represents the power of Additive White Gaussian Noise (AWGN). For the sake of simplicity, channel interference between nodes is ignored in this paper. Channel traversal rate is set as a constant value C, that is, users expect to receive all content blocks at the same rate, then the bandwidth consumption of base station node M can be written as:

$$B_m^{consu} = \sum_{l=1}^{L} max \frac{Cu_{iB_m}^l}{\log_2\left(1 + \frac{S_i^{B_m}}{N_i^{B_m}}\right)}, \forall i \in g^l. \tag{2}$$

Satellite Communication Model: For satellite communication mode, if user i requests content from satellite node r, the SNR is given by:

$$S_r^{SNR} = \frac{P_s G_t G_r \left(\frac{\lambda}{4\pi D}\right)^2}{d}}{k\mu}, \tag{3}$$

where P_s is the transmission power of satellite node, λ represents satellite wavelength, D is the altitude of LEO satellite, d is Atmospheric loss, k is the Boltzmann constant, μ is the noise temperature of the terminal, G_t and G_r are the typical antenna gains of the transmitter and receiver, respectively. In this paper, channel data rate is set to the same value C, and the bandwidth consumption of satellite-ground link of satellite node r is:

$$S_r^{consu} = \sum_{l=1}^{L} \varphi_{S_r}^l \cdot \frac{C}{\log_2\left(1 + \frac{\frac{P_s G_t G_r \left(\frac{\lambda}{4\pi D}\right)^2}{d}}{k\mu}\right)}, \tag{4}$$

where $\varphi_{S_r}^l = 1$ if $\sum_{l=1}^{L} u_{S_r}^l > 0$, otherwise $\varphi_{S_r}^l = 0$.

3.2 Problem Formulation

In this subsection, we construct the system utility model by considering the cache hit ratio and bandwidth consumption ratio to evaluate the SGIN system utility. The cache hit ratio is defined as the ratio between the number of user requests successfully satisfied and all of user requests, and the bandwidth consumption ratio is defined as the ratio between the bandwidth spectrum resources consumed by SGIN and the total budget. In this paper, the system utility of SGIN in a round of content delivery can be written as:

$$\mathbb{U} = r_1 g_1 - r_2 g_2, \ r_1 + r_2 = 1, \tag{5}$$

where r_x reflects the preference of different indicators, g_x represents the evaluation index after normalization.

The optimization object can be expressed as:

$$\max \mathbb{U} = r_1 g_1 - r_2 g_2, \tag{6}$$

subject to

$$u_{iB_m}^l + u_{iS_r}^l \le 1, \quad \forall i \in g^l, \forall l, m, r \tag{7}$$

$$u_{iB_m}^l \le x_{B_m}^l, \quad \forall i \in g^l, \forall l, m \tag{8}$$

$$u_{iS_r}^l \le \sum_{r=1}^{R} x_{S_r}^l, \quad \forall i \in g^l, \forall l \tag{9}$$

$$\sum_{l=1}^{L} x_{B_m}^l \le C_B, \quad \forall m \tag{10}$$

$$\sum_{l=1}^{L} x_{S_r}^l \le C_S, \quad \forall r \tag{11}$$

$$\sum_{l=1}^{L} \max \left(\frac{C u_{iB_m}^l}{\log_2 \left(1 + \frac{S_i^{B_m}}{N_i^{B_m}} \right)} \right) \le \mathbb{B}_1, \quad \forall i \in g^l, \forall m \tag{12}$$

$$\sum_{l=1}^{L} \varphi_{S_r}^l \cdot \frac{C}{\log_2 \left(1 + \frac{\frac{P_s G_t G_r}{d} \left(\frac{\lambda}{4 \pi D} \right)^2}{k \mu} \right)} < \mathbb{B}_2, \quad \forall r \tag{13}$$

$$u_{iB_m}^l \in \{0, 1\}, \quad \forall i \in g^l, \forall l, m \tag{14}$$

$$u_{iS_r}^l \in \{0, 1\}, \quad \forall i \in g^l, \forall l, r \tag{15}$$

$$x_{B_m}^l \in \{0, 1\}, \quad \forall l, m \tag{16}$$

$$x_{S_r}^l \in \{0, 1\}, \quad \forall l, r \tag{17}$$

$$\varphi_{S_r}^l \in \{0, 1\}, \quad \forall l, r. \tag{18}$$

g_1 and g_2 can be expressed as: $g_1 = \frac{1}{N} \sum_{l=1}^{L} \left(\sum_{r=1}^{R} \sum_{i=1}^{N} u_{iS_r}^l + \sum_{m=1}^{M} \sum_{i=1}^{N} u_{iB_m}^l \right)$ and

$$g_2 = \frac{C}{M \mathbb{B}_1 + R \mathbb{B}_2} \left(\sum_{m=1}^{M} \sum_{l=1}^{L} \max \frac{u_{iB_m}^l}{\log_2 \left(1 + \frac{S_i^{B_m}}{N_i^{B_m}} \right)} + \sum_{r=1}^{R} \sum_{l=1}^{L} \varphi_{S_r}^l \frac{1}{\log_2 \left(1 + \frac{\frac{P_s G_t G_r}{d} \left(\frac{\lambda}{4 \pi D} \right)^2}{k \mu} \right)} \right).$$

respectively. Constraint (7) ensures that users can only be severed by only one edge node. Constraint (8) indicates that an edge node can only serve the users requesting this content if it caches certain blocks of content. At the same time, constraint (9) indicates that the satellite nodes can share the cache through inter-satellite links. Constraints (10)–(13) are related to node storage resources and spectrum resources. In constraints (14)–(18), both $u^l_{iB_m}$ and $u^l_{iS_r}$ are 0–1 variables, which indicate that whether the user nodes i requests content from base station m or satellite node r. Both $x^l_{B_m}$ and $x^l_{S_r}$ are 0–1 variables, which indicate whether base station node M or satellite node R caches content block l. $\varphi^l_{S_r}$ is 0–1 variable, which indicates that whether provide service for a group of users of requesting content. It should be noted that the bandwidth consumed by a multicast user group is subject to the user with the worst channel in the multicast group.

4 JSSS Algorithm

In this paper, the optimization object is to maximize the system utility $\mathbb{U}$. We can see that g_1 and g_2 are subject to the user i, which belongs to a content group g^l. Obviously, the optimization of g_1 and g_2 is coupled, and the original problem is a mixed integer programming problem, belonging to the NP-hard problem. Therefore, the JSSS algorithm is proposed in this paper, as shown in Fig. 2. In order to decouple storage resources and spectrum resources, this algorithm divides the problem into two stages, namely CP stage and MT stage. In the CP stage, the MADRL based algorithm is proposed to learn the environment information and make the cache placement decision intelligently, aiming to improve the cache hit rate. In the MT stage, the genetic algorithm is proposed to match user nodes and edge nodes and form multicast groups, aiming to reduce the bandwidth consumption ratio. In the long term, the two stages are executed alternately to optimize the system utility.

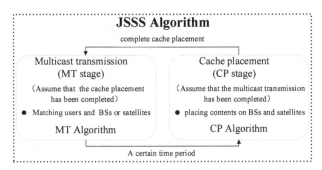

Fig. 2. The JSSS algorithm

4.1 MT Stage

In this subsection, the optimization of the MT stage in JSSS algorithm is introduced. In the MT stage, we assume that the cache placement (i.e., CP stage) has been completed, and we only need to optimize the bandwidth consumption ratio. Focusing on reducing the bandwidth consumption ratio, the problem of matching user and edge nodes in SGIN can be expressed as: how to match users and base station or satellite nodes to minimize the bandwidth consumption ratio with a given file popularity, node cache capacity and network topology? The optimization objective can be written as:

$$opt.\mathcal{A} = \min r_2 g_2, \tag{19}$$

subject to (7)–(18).

In order to prove that the optimization problem (19) is a NPC problem, we consider the corresponding decision problem of problem $\mathcal{A}$ in (20).

$$dec.\mathcal{B} = r_2 g_2 =$$

$$\frac{r_2 C}{M \mathbb{B}_1 + R \mathbb{B}_2} \left(\sum_{m=1}^{M} \sum_{l=1}^{L} \max \frac{u_{iBm}^l}{\log_2 \left(1 + \frac{S_i^{Bm}}{N_i^{Bm}}\right)} + \right.$$

$$\left. \sum_{r=1}^{R} \sum_{l=1}^{L} \varphi_{S_r}^l \frac{1}{\log_2 \left(1 + \frac{P_s G_t G_r}{d} \frac{\left(\frac{\lambda}{4\pi D}\right)^2}{k\mu}\right)} \right) \leq \varepsilon. \tag{20}$$

We prove that the problem $\mathcal{B}$ is a NPC problem with a special case, that is, if the special case of the problem $\mathcal{B}$ is a NPC problem, then so is the general case. The special case is given: 1) There are two base station nodes, one satellite node and one ground station node, i.e., $M = 2$, $R = 1$, $J = 1$; 2) Users 1 and 2 request content 1, and user 3 requests content 2. With other users joining, the channel condition is worse than that of existing users; 3) Cache capacity of each node is 1, i.e., $C_B = 1, C_S = 1$; 4) The spectrum resources of SGIN are sufficient to complete the content distribution for three users; 5) Base station node 1 serves for the user group 1, satellite node 1 serves for the user group 2 and the bandwidth consumption ratio is as ε.

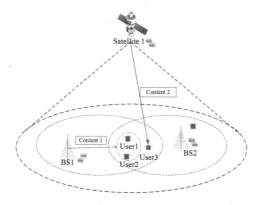

Fig. 3. The equivalent explanation of problem $\mathcal{B}$ and MWSCP

Lemma 1: The solve of problem $\mathcal{B}$ is equivalent to solve the minimum weighted set covering problem (MWSCP).

Proof: Figure 3 shows the equivalence of the original problem $\mathcal{B}$ and the MWSCP problem. If we assign user group 1 containing user 1 and user 2 to base station node 1 and user group 2 containing user 3 to satellite node 1, the problem $\mathcal{B}$ can be solved. Thus, Lemma 1 is proved.

Since the MWSCP is a NPC problem, the problem $\mathcal{B}$ is a NPC problem according to Lemma 1. The problem $\mathcal{B}$ is a corresponding decision form of problem $\mathcal{A}$, therefore problem $\mathcal{A}$ is also a NPC problem. The optimization of problem $\mathcal{A}$ is transformed into bipartite graph in order to simplify the original problem. Denote $F = \{F_p | p \in \mathbb{N}^+\}$ as a user set requesting the same content and accessing to the same edge node. Problem $\mathcal{A}$ is presented as $<U, F>$. Let each element of U be the vertex set U of a bipartite graph, and let each element of F (i.e., a group of users) be another vertex set F of a bipartite graph. If the element F_p in the set F containing the element U_i in the set U, we connect into a line between element F_p and element U_i. Then, we get a graph $<U, E>$, $V = U \cup F$ and $E = \{(u, f) | u \in U, f \in F, u \in f\}$. Thus problem $\mathcal{A}$ is converted to the following problem: Find a subset F^* with minimal weight in the vertices set F, where every vertex in U has at least one edge connected to some vertex in F^*.

Lemma 2: If the node V is covered only by the set F_p, then F_p is in F.

Proof: In F, all nodes should be covered, and node V is covered only by the set F_p. Therefore, if node V should be covered, then F_p should be in F. Thus, Lemma 2 is proved.

According to Lemma 2, users who are not within the communication range of the base station node or whose requested content is not cached in the base station node can be divided into different user groups according to the requested content. These users denote as $\tilde{U}$, whose request can only be satisfied by satellite nodes. Therefore, in order to simplify the original problem, the users in the set of $\tilde{U}$ are deleted from the user set U, and we only need to consider users accessing to contents from base station node (i.e., $\dot{U}$). It is worth noting that, the users without the service will be further merged into the set $\tilde{U}$ when match the users and base station nodes.

Lemma 3: For any two subsets F1 and F2 in set F, the subset F2 and its associated edges can be deleted from the bipartite graph if their neighboring node set satisfies the condition $\xi(F1) \subseteq \xi(F2)$ and its weight satisfies $\omega(F_1) \leq \omega(F_2)$.

Proof: Since the $F2$ can be completely replaced by $F1$, and the total cost will be lower.

According to Lemma 1, the original problem belongs to MWSCP problems, and the heuristic solution strategy that can reduce the cost of computing resources is more suitable for solving the original problem. Then, the original problem is equivalent transformed by bipartite graph theory. Using the properties of bipartite graph (i.e., Lemma 2 and Lemma 3), we simplify the original problem and the solving complexity is reduced. Further, we put forward the heuristic algorithm based on genetic algorithm to match $\dot{U}$ with the base station node. In a word, the matching in the MT stage is shown in Fig. 4.

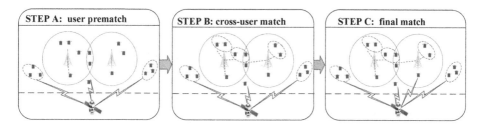

Fig. 4. The workflow diagram of the MT stage

4.2 CP Stage

In this subsection, the optimization of the CP stage in JSSS algorithm is introduced. In the CP stage, we assume that the multicast transmission (i.e., MT stage) has been completed, and we only need to optimize the cache hit ratio. Aiming at improving cache hit ratio, the cache placement problem in SGIN can be expressed as: how to cache content on base station and satellite nodes to maximize cache hit rate for given user request history, cache capacity of nodes and network topology, etc. The optimization object can be written as

$$opt.\mathcal{C} = \max r_1 g_1, \tag{21}$$

subject to (7)–(18).

In reality, the change of content popularity is complex and unstable due to the dynamic changing demand of content. In this case, machine learning techniques can be used to learn user's preferences based on historical information and determine which content to cache on the MEC server. However, in the context of the content distribution in SGIN, the process of cache placement involves interactions between multiple cache nodes. In order to solve the optimization problem $\mathcal{C}$, we design the content caching scheme of MEC server by utilizing MADRL.

Figure 5 shows the application of DRL in an edge computing enabled SGIN. Satellite nodes use global information in the process of centralized training. However, each edge node uses local information to make cache decisions in distributed execution. In Fig. 5, the caching proxy gets some raw data (such as user requests, network conditions, and historical user settings, etc.) by observing the state of the environment. Deep Neural Networks (DNNs) receives this data and outputs value functions or actions. Based on the result of output, the proxy selects a caching action and observes the rewards result of corresponding action. Then, the agent can use rewards to train and improve the deep neural network model.

As shown in Fig. 6, the proposed algorithm includes two neural networks (i.e., actor network and critic network). x_l^i represents the number of requesting content l in the i-th circle, and y_k^i represents the set of users in the i-th circle. In Fig. 6, the advantage update is combined with the actor and critic network, which trains and updates in an asynchronous and parallel pattern. DRL training is performed on each base station/satellite node, and we can predict content popularity of the next round and cache content based on user history request patterns, node cache hit ratio and bandwidth consumed ratio in the coverage of

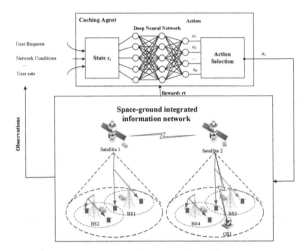

Fig. 5. The application of DRL in an edge computing enabled SGIN

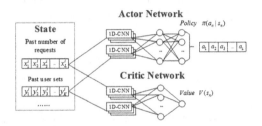

Fig. 6. The actor and critic strategy in the CP stage

base station/satellite. Then, the cache probability of each content is obtained through the actor network and the cache capacity of all nodes is updated. Finally, the MT algorithm is executed based on the output of the CP stage. Obviously, the above operations are performed independently and asynchronously between each node.

5   Simulation Results and Analysis

In this section, in order to verify the efficiency of the algorithm, we use real MovieLens data (including 138000 users' 20 million scores on 27000 movies), which can be obtained from https://grouplens.org/datasets/movielens. Due to the fact that the access pattern displayed by the scored data is similar to the pattern of content request, we assume that the score of each movie is a video request.

In addition, the network topology and channel parameters in each round will change with the distribution of users and the content of request. In the CP stage, the satellite nodes directly issue the cache decision to each edge node in the training. After updating the DRL parameters, all edge nodes make the cache decision in the execution, while the DRL model parameters are periodically updated and issued by the satellite node. Simulation parameters are set as Table 1.

Table 1. Simulation parameters

Simulation parameters	Value
The altitude of satellite	800 km
The number of users	100
The number of contents	50
The spectral bandwidth of $\mathbb{B}_1$	20 MHz
The spectral bandwidth of $\mathbb{B}_2$	30 MHz
The noise power spectral density	−174 dBm/Hz
The transmission power	43 dBm
The noise temperature of terminal	290 K
The antenna gains of transmitter	54.4 dBw
The antenna gains of receiver	0 dB

5.1 Performance of SGIN and Terrestrial Network

Figures 7, 8, and 9 show the system utility results comparison between SGIN and terrestrial network as the round increases. As fair as possible, both terrestrial and SGIN networks adopt JSSS algorithm, and the difference is that only the terrestrial controller is used in the terrestrial network to obtain the global information for training DRL model in the CP stage, without caching and distributing content. The first 900 rounds are in the centralized training stage, while the remaining 100 rounds are in the distributed execution stage. System utility is a weighted value of bandwidth consumption ratio and cache hit ratio. It can be seen from the simulation results that SGIN has higher cache hit ratio, lower bandwidth consumption ratio and higher system utility than terrestrial network. Figure 7 shows that the system utility of SGIN network outperforms terrestrial network by about 15.4% when the number of rounds is equal to 900. Figure 8 shows that the bandwidth consumption ratio of SGIN is lower about 20.0% compared to the terrestrial network when the number of rounds is 900. Figure 9 shows that the cache hit ratio of SGIN outperforms terrestrial network by about 13.6% when the number of rounds is 900. Therefore, it can be proved that the proposed SGIN scheme can significantly improve the system performance.

5.2 Performance of Different Matching Algorithms in the MT Stage

Figure 10 shows the system utility of different matching algorithms as the round increases in the MT stage. We verify the performance of our scheme by comparing MT algorithm with other three algorithms, namely random selection algorithm, greedy algorithm and optimum algorithm. As fair as possible, the four algorithms are carried out in the same way as JSSS in the CP stage. Random selection algorithm refers to users and edge nodes randomly matched in the MT stage; Greedy algorithm refers to greedy matching between users and edge nodes in the MT stage. The optimum algorithm refers to the exhaustive matching of users and edge nodes in the MT stage. As can be seen from Fig. 10, when the

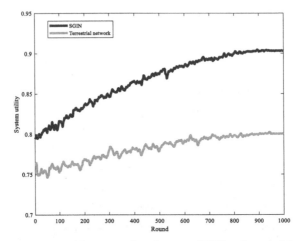

Fig. 7. The system utility comparison between SGIN and terrestrial network

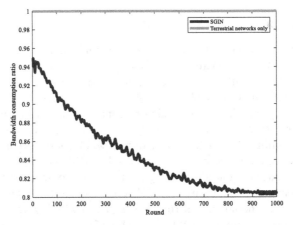

Fig. 8. The bandwidth consumption ratio comparison between SGIN and terrestrial network

nodes and users are matched in the MT algorithm, the system utility is higher than that of the greedy algorithm. For example, the system utility of MT algorithm outperforms with the greedy algorithm by about 9.8% when the number of rounds is equal to 900, due to edge nodes and users matched to achieve local optimality in greedy algorithm. In addition, the performance gap between the MT algorithm and the optimum algorithm is very small. When the number of rounds is equal to 900, the MT algorithm only reduces the system utility by 2.2%.

5.3 Performance of Different Caching Algorithms in the CP Stage

Figure 11 shows the system utility of different cache placement algorithms as the round increases in the CP stage. The performance of the proposed algorithm is verified by comparing the CP algorithm with other three cache placement algorithms, namely fixed

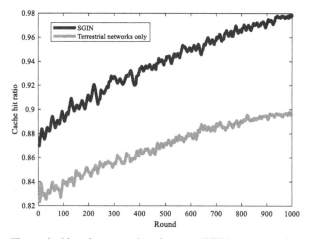

Fig. 9. The cache hit ratio comparison between SGIN and terrestrial network

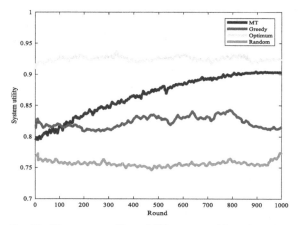

Fig. 10. The system utility of different matching algorithms

cache, random cache and the popularity-based cache algorithm. As fair as possible, the four algorithms are carried out in the same way as JSSS in the MT stage. The fixed cache algorithm means that all nodes cache the same content all the time during the CP stage (the initial contents are randomly placed); Random cache algorithm means that nodes cache contents randomly in each round of the CP stage; The popularity-based cache algorithm means that all nodes cache the most popular content in each round of the CP phase. Figure 11 shows that the system utility of proposed CP algorithm outperforms with other algorithms when the number of rounds is equal to 900. The reason is that users may request different contents in different rounds, and the user requests in some areas may show patterns that are difficult to detect without historical data. By analyzing the historical user requests recorded in nodes, the proposed scheme can discover the potential patterns of users.

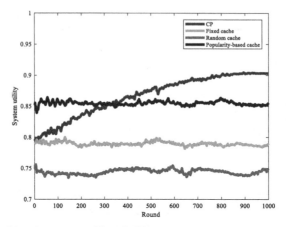

Fig. 11. The system utility of different cache placement algorithms

Figure 12 shows the system utility of multi-agent and single-agent as the round increases in the CP stage. As can be seen from Fig. 12, the system utility of multiple agents outperforms with the single agent by about 9.2% when the number of rounds is equal to 900. This is due to that the satellite node of the multi-agent algorithm can obtain additional information in the centralized training stage, while the node of the single-agent algorithm can only obtain local information.

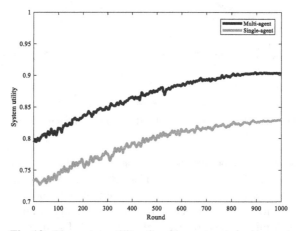

Fig. 12. The system utility of multi-agent and single-agent

5.4 Performance of Different Base Station Node Parameters

Figure 13 shows the system utility of different cache capacity as the round increases. The cache capacity is taken as 10, 15 and 20 blocks, respectively. With the increase of cache capacity, the system utility also increases. The reason is that the cache hit ratio

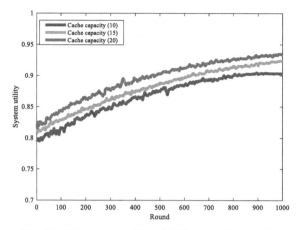

Fig. 13. The system utility of different cache capacity

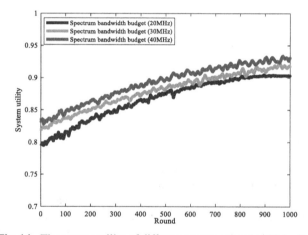

Fig. 14. The system utility of different spectrum bandwidth budget

increases with the increase of the cache capacity, and the bandwidth consumption ratio decreases with the increase of the cache capacity, thus improving the system utility. Figure 14 shows the system utility of different spectrum bandwidth budget as the round increases, and the value of spectrum bandwidth is 20, 30 and 40 MHz, respectively. With the increase of the base station spectrum bandwidth budget, the system utility also increases. The reason is that the bandwidth consumption ratio decreases with the increase of the spectrum bandwidth budget, and the cache hit ratio increases with the increase of the spectrum bandwidth budget, thus improving the system utility. Figure 15 shows the system utility of different number of served users as the round increases. With the increase of the number of users served by a single base station node, the system utility decreases accordingly. The reason is that the bandwidth consumption ratio increases with the increase of the number of users served by the single base station node, while the cache hit ratio decreases with the increase of the number of users served by the single base station node. Extensive simulations prove that the proposed JSSS is reasonable, which can effectively improve the cache hit ratio, reduce the consumption of spectrum resources and improve the system utility.

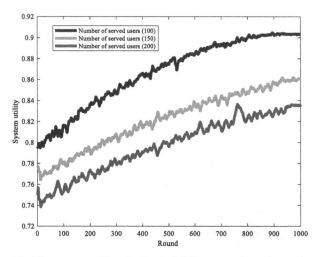

Fig. 15. The system utility obtained of different number of served users

6   Conclusion

In this paper, we proposed a JSSS algorithm for the joint scheduling of communication resources and storage resources to maximize the system utility in an edge computing enabled SGIN. We solved the optimization problem with a phased strategy. In the MT stage, we matched users and BSs or satellites based on a heuristic genetic algorithm. In the CP stage, we utilized a MADRL-based algorithm to place contents adaptively. Simulations showed that the proposed scheme can achieve superior performance.

References

1. Thota, J., Bulut, B., Doufexi, A., Armour, S.: Performance evaluation of multicast video distribution with user cooperation in LTE-A vehicular environments. In: IEEE 86th Vehicular Technology Conference (VTC-Fall), Toronto, pp. 1–5 (2017)
2. Cisco visual networking index: forecast and trends, 2017–2022 white paper [EB/OL]. https://www.cisco.com/c/en/us/solutions/collateral/service-provider/visual-networking-index-vni/white-paper-c11-741490.html#_Toc532256803
3. Vergel, R.S., Tena, P.M., Yrurzum, S.C., Cruz-Neira, C.: A comparative evaluation of a virtual reality table and a hololens-based augmented reality system for anatomy training. IEEE Trans. Hum. Mach. Syst. **50**(4), 337–348 (2020)
4. Tao, X., Duan, Y., Xu, M., Meng, Z., Lu, J.: Learning QoE of mobile video transmission with deep neural network: a data-driven approach. IEEE J. Sel. Areas Commun. **37**(6), 1337–1348 (2019)

5. Tao, X., Chen, Z., Xu, M., Lu, J.: Rebuffering optimization for DASH via pricing and EEG-Based QoE modeling. IEEE J. Sel. Areas Commun. **37**(7), 1549–1565 (2019)
6. Abbas, N., Zhang, Y., Taherkordi, A., Skeie, T.: Mobile edge computing: a survey. IEEE Internet of Things J. **5**(1), 450–465 (2018)
7. Tu, Z., Zhou, H., Li, K., Li, G., Shen, Q.: A routing optimization method for software-defined SGIN based on deep reinforcement learning. In: IEEE Globecom Workshops (GC Wkshps), Waikoloa, HI, USA, pp. 1–6 (2019)
8. OneWeb makes history as first launch mission is a success [EB/OL]. https://www.oneweb.net/newsroom/oneweb-makes-history-as-first-launch-mission-is-a-success
9. SpaceX Starlink: Here's everything you need to know [EB/OL]. https://www.digitaltrends.com/cool-tech/what-is-spacex-starlink/
10. Zhu, X., Jiang, C., Yin, L., Kuang, L., Ge, N., Lu, J.: Cooperative multigroup multicast transmission in integrated terrestrial-satellite networks. IEEE J. Sel. Areas Commun. **36**(5), 981–992 (2018)
11. Yan, S., Qi, L., Peng, M.: User access mode selection in satellite-aerial based emergency communication networks. In: IEEE International Conference on Communications Workshops (ICC Workshops), Kansas City, MO, pp. 1–6 (2018)
12. Wu, H., Li, J., Lu, H., Hong, P.: A two-layer caching model for content delivery services in satellite-terrestrial networks. In: IEEE Global Communications Conference (GLOBECOM), Washington, DC, pp. 1–6 (2016)
13. Liu, S., Hu, X., Wang, Y., Cui, G., Wang, W.: Distributed Caching Based On Matching Game in LEO satellite constellation networks. IEEE Commun. Lett. **22**(2), 300–303 (2018)
14. Zhong, G., Yan, J., Kuang, L.: QoE-driven social aware caching placement for terrestrial-satellite networks. China Commun. **15**(10), 60–72 (2018)
15. Zhu, H., Cao, Y., Wei, X., Wang, W., Jiang, T., Jin, S.: Caching transient data for Internet of Things: a deep reinforcement learning approach. IEEE Internet of Things J. **6**(2), 2074–2083 (2019)
16. Zhu, H., Cao, Y., Wang, W., Jiang, T., Jin, S.: Deep reinforcement learning for mobile edge caching: review, new features, and open issues. IEEE Netw. **32**(6), 50–57 (2018)
17. Jiang, M., Hai, T., Pan, Z., Wang, H., Jia, Y., Deng, C.: Multi-agent deep reinforcement learning for multi-object tracker. IEEE Access **7**, 32400–32407 (2019)

Dynamic Priority-Based Computation Offloading for Integrated Maritime-Satellite Mobile Networks

Ailing Xiao[1(✉)], Haoting Chen[1], Sheng Wu[2], Li Ma[1], Fan Zhou[3], and Dongchao Ma[1]

[1] School of Information Science and Technology, North China University of Technology,
Beijing 100144, China
xiaoailing@tsinghua.edu.cn

[2] School of Information and Communication Engineering, Beijing University of Posts and
Telecommunications, Beijing 100876, China

[3] School of Information Science and Technology,
Shenyang Ligong University, Shenyang 110159, Liaoning, China

Abstract. In order to meet the increasing demand for delay-sensitive and computing-intensive applications of maritime users, we first propose an integrated satellite-maritime mobile edge computing framework. The framework considers a maritime mobile communication network consisting of maritime satellites and shipborne base stations, and maritime computation offloading coordinated by the terrestrial cloud and the shipborne edge servers. Secondly, we dynamically calculate the priority that characterizes the urgency of offloading tasks with reinforcement learning. Based on the dynamic priority, a maritime computation offloading method is proposed to optimize the system cost. Finally, simulation results verify the effectiveness and convergence of the proposed method in terms of offloading delay, user energy consumption, offloading response rate and average offloading cost.

Keywords: Integrated maritime-satellite mobile networks · Mobile edge computing · Reinforcement learning · Dynamic priority

1 Introduction

With the continuous development of the maritime economy, maritime activities have become more frequent. Marine applications of Internet of things technology are also increasing, such as maritime remote sensing buoy sensor, unmanned aerial vehicle autopilot, maritime mobile video analysis, etc. [1, 2]. The great demands of delay-sensitive computing-intensive intelligent applications by maritime users could consume lots of computing resources, thus require the support of the computation offloading service [3]. As a supplement to the cloud computing, mobile edge computing (MEC)

Foundation Items: National Nature Science Foundation of China (62001007, 62022019); Start-up Fund for Newly Introduced Teacher (110051360002).

© Springer Nature Singapore Pte Ltd. 2021
Q. Yu (Ed.): SINC 2020, CCIS 1353, pp. 70–83, 2021.
https://doi.org/10.1007/978-981-16-1967-0_5

can provide computation offloading for nearby users, which has the advantages of low latency and low energy consumption. In recent years, MEC has been widely concerned by both academia and industry [4]. In order to minimize the offloading delay and energy consumption of users within a single cell, Chen et al. [5] proposed an edge computation offloading method based on the relay server, which allowed multiple users to share task processing results. Wang et al. [6] proposed a pattern recognition algorithm for task offloading with deep learning, which determined whether to offload the task to the edge server or at the user end through automatic recognition of the task pattern.

However, when the resource needed by the offloading tasks exceeds the processing capacity of the edge node (EN), edge computing alone cannot meet the offload requirements. Cloud computing and edge computing each have their own values. We should make collaborative use of both the cloud and the edge resources to minimize the offloading cost and relieve the pressure on network resources [7–10]. In this paper, we consider the application of the cloud-edge collaborative computation offloading in the maritime mobile network. Du et al. [7] studied the user computation offloading of a hybrid fog/cloud system and ensured the fairness of computation offloading among users by balancing the offloading costs of all users. Huang et al. [8] implemented differential offloading for tasks with different delay requirements, and used cloud resources and edge resources to minimize the offloading delay. However, the existing studies have not fully considered the dynamic changes of the emergency degree of tasks under dynamic network conditions, which may lead to excessive allocation of computing resources to non-urgent tasks and affect the timely response rate.

Cloud-edge collaboration is inseparable from the support of communication networks. Traditional maritime broadband communication methods include WLAN, Wi-Fi, onshore base station, and maritime satellite [11]. The coverage of WLAN, Wi-Fi, and the onshore base station are limited, although the maritime satellite can provide broadband communication anytime and anywhere, its communication price is high. Combined with the advantages of traditional maritime broadband communication, "integrated maritime communication network architecture" [12, 13] tries to provide broadband access with moderate price and wide area coverage for maritime mobile users. Based on wireless Mesh technology, projects like BLUECOM+ [12] make use of the mesh nodes installed on tethered balloons and UAVs to form a maritime mesh network. With the help of onshore base stations, it can provide broadband coverage for 100 km of coastal waters, with an average data rate of more than 3 Mbps. With the cooperation of coastal base stations and shipborne LTE routers, LTE-Maritime [13] can achieve the same maritime broadband coverage as BLUECOM+. "Integrated maritime communication network" is suitable as the infrastructure of maritime cloud-edge offloading.

In summary, combined with shipborne base station (S-BS) and MEC, this paper first proposes an integrated satellite-maritime mobile edge computing framework (ISM-MEC), which considers the deployment of base stations and edge servers on large ships to provide edge computing services for the surrounding ships and users. With ISM-MEC, the reinforcement learning method is used to dynamically output the priority of the offloading tasks according to the network conditions, and the dynamic priority-based computation offloading (DPCO) is proposed for the maritime offloading task. The simulation results show that the proposed method can reduce the delay of maritime computation offloading tasks and the energy consumption of user ends, and maintain a high timely response rate of offloading tasks.

2 System Model

This section first introduces the integrated satellite-maritime mobile edge computing framework, then introduces the communication model and computing model in the architecture, and models the performance indicators of cloud edge offload, including offload delay, user terminal energy consumption, and offloading task response rate. Finally, modeling the maritime computation offloading problem.

2.1 Integrated Satellite-Maritime Mobile Edge Computing Framework

As shown in Fig. 1, the integrated satellite-maritime mobile edge computing framework in this paper mainly includes four layers: (1) The terrestrial cloud service layer is composed of CS (cloud server), (2) The satellite link layer consists of broadband satellite and narrowband satellite, (3) The maritime edge layer is composed of edge node ships, on which satellite communication equipment, shipborne base station, and edge server are deployed, (4) The maritime user layer is composed of the maritime mobile user equipment (UE) around the edge node ship. The EN ship has computing resources, and a variety of delay-sensitive computing-intensive applications run on the UE. When the UE needs to offload, it can offload the computing tasks to the EN ship providing broadband signal nearby, if the offloading task exceeds the computing capacity of the EN ship, the EN ship can offload the computing tasks to the terrestrial cloud server through its satellite communication.

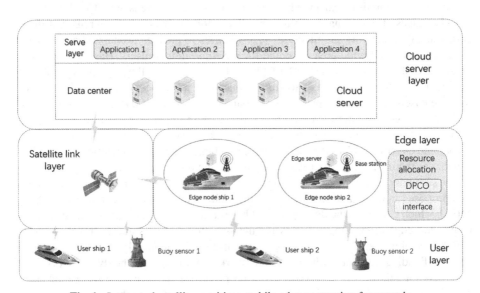

Fig. 1. Integrated satellite-maritime mobile edge computing framework.

Table 1 lists the symbols used in this paper and their definition. At time t, we consider an maritime mobile network composed of $N(t)$ EN ships and $M(t)$ UEs. Where the set of UE is $U(t) = \left\{ U_j^t | j = 1, \ldots, M(t) \right\}$, The set of EN ships is $B(t) = \{ B_i^t | i = 1, \ldots, N(t) \}$. The number of each B_i^t communication resource block (RB, resource block) is R_{max}, and is evenly allocated to the UE set $S_i(t)$ it serves. The total amount of computing resources for each B_i^t is F_{max}, and each U_j^t occupies at least one RB, $|S_i(t)| \leq R_{max}$.

Table 1. Symbol summary.

Symbol	Definition
U_j^t	User ship
B_i^t	Edge node ship
F_{max}	Computing resource capacity of edge nodes
$f_{i,j}^e(t)$	The computing resources of B_i^t assign to U_j^t
$s_{j,i}^e(t)$	Whether U_j^t offload task to B_i^t
$s_j^c(t)$	Whether U_j^t offload task to CS
$Cost_j(t)$	The weighted sum of cost, delay and energy consumption of task offloading
$\Gamma_{j,i}(t)$	The signal to noise ratio of U_j^t and B_i^t
t_{thr}	The threshold of delay tolerance for offloading tasks

2.2 Communication Model

When U_j^t was served by B_i^t, the transmitting power of U_j^t is $Pt(j, i, t)$, the transmitting power of B_i^t is $Pt^b(i, j, t)$, the receive power of U_j^t is:

$$Pr^b(j, i, t) = 10 \, log_{10}\left(Pt^b(i, j, t)\right) - L_{j,i}(t) \tag{1}$$

Where $L_{j,i}(t)$ is the estimation of path loss between B_i^t and U_j^t, the following two ray model is used for estimation:

$$L_{j,i}(t) = -10 \, log_{10}\left\{ \left(\frac{\lambda}{4\pi d_{j,i}} \right)^2 \left[2 \sin\left(\frac{2\pi h_i h_j}{\lambda d_{j,i}} \right) \right]^2 \right\} \tag{2}$$

Where λ is the wavelength, h_i is the height of B_i^t, h_j is the height of U_j^t, $d_{j,i}$ is the distance between B_i^t and U_j^t.

The achievable downlink rate of U_j^t is:

$$R^b(j, i, t) = Bs_{i,j} \cdot log_2\left(1 + \Gamma_{i,j}^b(t)\right) \tag{3}$$

Where $Bs_{i,j}$ is the bandwidth of B_i^t allocated to U_j^t, $\Gamma_{i,j}^b(t)$ is the signal to noise ration of B_i^t receiving data:

$$\Gamma_{i,j}^b(t) = \frac{Pr^b(j,i,t)}{\sigma + \sum_{n=1,n\neq i}^{N(t)} a_n(t)Pr^b(n,i,t)} \tag{4}$$

Where σ is the noise power, $\sum_{n=1,n\neq i}^{N(t)} Pr^b(n, i, t)$ indicates the interference power caused by the surrounding UEs.

The receiving power of B_i^t is:

$$Pr(i, j, t) = 10\, log_{10}(Pt(j, i, t)) - L_{j,i}(t) \tag{5}$$

The uplink rate of U_j^t is:

$$R(i, j, t) = Bs_{i,j} \cdot log_2\left(1 + \Gamma_{j,i}(t)\right) \tag{6}$$

The signal to noise ratio when U_j^t receiving data is:

$$\Gamma_{j,i}(t) = \frac{Pr(i,j,t)}{\sigma + \sum_{n=1,n\neq i}^{N(t)} Pr^b(n,i,t)} \tag{7}$$

2.3   Computation Model

We assume U_j^t generate computing tasks $\theta_j(t)$, the amount of data is $b(\theta_j(t))$, the CPU cycle required to complete the task is $d(\theta_j(t))$, the maximum tolerance delay is $delay^{max}(\theta_j(t))$, the data volume of the processing result is $\beta(\theta_j(t))$. There are three alternative computing models: local computing, maritime edge computing and terrestrial cloud computing, we model the offloading cost of the three computing modes from two aspects of offloading delay and energy consumption.

For local computing, it is assumed that the computing frequency of UE (i.e., CPU cycles per second) is $f_j^u(t)$. Then the local computation delay is:

$$t_j^u(t) = \frac{d(\theta_j(t))}{f_j^u(t)} \tag{8}$$

K_j is the energy consumption per unit task cycle, the user energy consumption calculated locally is:

$$e_j^u(t) = K_j d(\theta_j(t)) \tag{9}$$

Therefore, the offloading cost of local computing is calculated by:

$$C_j^u(t) = \alpha_t t_j^u(t) + \alpha_e e_j^u(t) \tag{10}$$

Where α_t and α_e represent delay and energy consumption, respectively $\alpha_t + \alpha_e = 1$.

For maritime edge computing, $\theta_j(t)$ will be transmitted through the shipboard base station to B_i^t, compared with local computing, U_j^t will bear additional costs in terms of delay and energy consumption. Suppose that the CPU frequency assigned to $\theta_j(t)$ is $f_{i,j}^e(t) \in [0, F_{max}]$, according to the communication model, we can get the delay of maritime edge computation as follows:

$$t_{i,j}^e(t) = \frac{b(\theta_j(t))}{R(i,j,t)} + \frac{d(\theta_j(t))}{f_{i,j}^e(t)} + \frac{\beta(\theta_j(t))}{R^b(j,i,t)} \tag{11}$$

Assuming that the receiving power of the base station is minimal and can be ignored. The user energy consumption of maritime edge computing is calculated by

$$e_{i,j}^e(t) = \left(Pt(j,i,t) + Pr^b(j,i,t)\right)\frac{b(\theta_j(t))}{R(i,j,t)} \tag{12}$$

Therefore, the offloading cost of maritime edge computing is:

$$C_{i,j}^e(t) = \alpha_t t_{i,j}^e(t) + \alpha_e e_{i,j}^e(t) \tag{13}$$

For terrestrial cloud computing, U_j^t will use satellite communication through B_i^t to offload $\theta_j(t)$ to the terrestrial cloud server. Suppose the computing frequency of the terrestrial cloud server is $f_j^c(t)$, the round-trip time of satellite link is rtt. According to the communication model, we can get the delay of terrestrial cloud computing as follows:

$$t_j^c(t) = \frac{b(\theta_j(t))}{R(i,j,t)} + \frac{d(\theta_j(t))}{f_j^c(t)} + \frac{\beta(\theta_j(t))}{R^b(j,i,t)} + rtt \tag{14}$$

The user energy consumption of terrestrial cloud computing is as follows:

$$e_j^c(t) = Pt(j,i,t)\frac{b(\theta_j(t))}{R(i,j,t)} \tag{15}$$

Therefore, the offloading cost of terrestrial cloud computing is calculated by:

$$C_j^c(t) = \alpha_t t_j^c(t) + \alpha_e e_j^c(t) \tag{16}$$

Within long-term T, the offloading delay is lower than $delay^{max}(\theta_j(t))$ is N^s and the total number of tasks is N^T, the response rate of offloading task is:

$$R^s = \frac{N^s}{N^T} \tag{17}$$

In addition, the variable is defined as $f_{i,j}^e(t) \in \{f_1, f_2\}$, it is the computing resource assigned to U_j^t, f_1 is the default allocation, f_2 is as twice as f_1.

2.4 Problem Formulation of Maritime Computation Offloading

The above three offloading costs are expressed as follows:

$$Cost_j(t) = \begin{cases} C_j^c(t), & \text{If the task is offloaded to the terrestrial CS} \\ C_{i,j}^e(t), & \text{If the task is offloaded to } B_i^t, \\ C_j^u(t), & \text{If the task is computed locally.} \end{cases} \tag{18}$$

Minimizing the offloading cost of maritime computing can be modeled as an integer linear programming problem

Variables:

$$s^e_{j,i}(t), s^c_j(t), f^e_{i,j}(t)$$

Objective function:

$$\text{Min: } \sum_{t=1}^{T} \sum_{j=1}^{M(t)} Cost_j(t) \tag{19}$$

Subject to:

$$C1: \sum_{j=1}^{M(t)} s^e_{j,i}(t) \leq R_{max}, \forall i, t \tag{20}$$

$$C2: \sum_{j=1}^{M(t)} f^e_{j,i}(t) \leq F_{max}, \forall i, t \tag{21}$$

$$C3: \sum_{j=1}^{M(t)} \sum_{i=1}^{N(t)} s^e_{j,i}(t) t^e_{i,j}(t) + \sum_{j=1}^{M(t)} s^c_j(t) t^c_j(t) \leq \text{delay}^{max}(\theta_j(t)), \forall i, t \tag{22}$$

$$C4: s^e_{j,i}(t) \in \{0, 1\}, s^c_j(t) \in \{0, 1\}, \forall j, i, t \tag{23}$$

$$C5: s^e_{j,i}(t) + s^c_j(t) = 1 \tag{24}$$

$$C6: f^e_{i,j}(t) \in \{f_1, f_2\} \tag{25}$$

Where C1 represents each U^t_j occupies at least one RB, and each B^t_i can only serve R_{max} users, C2 means that the computing resources allocated by B^t_i to U^t_j cannot exceed its maximum computing capacity, C3 indicates that the computing task of cloud edge collaborative offloading should be completed within the maximum tolerable delay, $s^e_{j,i}(t)$ and $s^c_j(t)$ are two binary variables that indicate the offloading selection of U^t_j, $s^e_{j,i}(t)$ indicates whether to offload to maritime edge node, and 1 indicates to offload to maritime edge node, $s^c_j(t)$ indicates whether to offload to terrestrial cloud server, and 1 indicates to offload to cloud server. C4 limited the range of $s^e_{j,i}(t)$, $s^c_j(t)$. C5 limits a single task can only choose one of cloud offloading and edge offloading, and C6 limits the value range of computing resource types.

3   Maritime Computation Offloading Method Based on Dynamic Priority

This section introduces the dynamic priority-based maritime computation offloading method under the maritime mobile edge computing architecture based on an integrated satellite-maritime mobile edge computing framework and describes the process of optimizing maritime computation offloading using the reinforcement learning optimization method.

Algorithm 1 Maritime computation offloading method based on dynamic priority

1) Initialization: Agent output Γ_{thr}, t_{thr}
2) while tasks has not been traversed
3)   if $delay^{max}(\theta_j(t)) > t_{thr}$
4) $\theta_j(t)$ is non-urgent task
5)   else
6) $\theta_j(t)$ is urgent task
7)   end if
8) end while
9) while UEs has not been traversed
10)   while ENs has not been traversed
11)   Find the ship with the highest $\Gamma_{i,j}^b(t)$ around
12)   if $\Gamma_{i,j}^b(t) > \Gamma_{thr}$
13) $s_{j,i}^e(t) = 1$
14)   else
15) $s_{j,i}^e(t) = 0$
16)   end if
17)   end while
18) end while
19) while ENs has not been traversed
20)   The computing resources are assigned to the urgent tasks, and the $f_{i,j}^e(t)$为f_2
21)   The $f_{i,j}^e(t)$ of non-urgent tasks is f_2
22)   if EN lack of computing resources
23) $s_j^c(t) = 1$
24)   end if
25) end while

The offloading strategy has an important impact on the performance of computing. In the integrated satellite-maritime mobile edge computing framework, the offloading strategy should be properly formulated. Different offloading tasks have different delay requirements. Under the condition of limited computing resources, the computing tasks can be divided into urgent tasks and non-urgent tasks, and the urgent tasks are given priority to allocate high-quality resources. We set the system parameter (Γ_{thr}, t_{thr}), Γ_{thr} is the signal-to-noise ratio threshold, which is related to task allocation (i.e., user association). t_{thr} is the tolerable delay threshold of offloading task, which is related to the judgment of task urgency. Signal to noise ratio (SNR) is an important performance index of communication quality and reliability. When maritime users are offloading, they must ensure that SNR is good enough. Using Γ_{thr}, UEs can access the base station with good SNR. Task tolerable delay is a key requirement of computing intensive tasks, and tasks must be completed within tolerable delay. Using t_{thr}, can let the urgent task get more computing resources, ensure that the task can be completed in time. Previous studies use fixed threshold to determine the priority of the offloading task. However, under the condition of dynamic topology of maritime mobile network, the urgency of offloading task also changes dynamically, and the use of fixed threshold will affect the reasonable allocation of computing resources. Therefore, this paper considers maritime computation offloading based on dynamic priority. As shown in Algorithm 1, after the priority of offloading tasks is obtained, the EN ship with the largest signal-to-noise ratio is found in turn. If the computing resources of the EN ship are sufficient, the computing resources are allocated, and the tasks are processed according to the urgency of the task. Otherwise, the EN ship is offloaded to the cloud server. If there is no EN ship that can be accessed around UE, only local computing can be used. However, maritime computation offloading based on dynamic priority is a nonconvex integer programming. When the number of offloading tasks is large, traditional methods are difficult to solve. In this paper, the reinforcement learning method is used to generate the dynamic priority, and then Algorithm 1 is used to complete the maritime computation offloading.

In this paper, the reinforcement learning method is used to analyze the system parameters (Γ_{thr}, t_{thr}) to minimize the cost of maritime computation offloading. Where, Γ_{thr} is the signal-to-noise ratio of users, which is related to the access of maritime mobile users. t_{thr} is the tolerable delay threshold of offloading task, which is related to the judgment of task's urgency. The overall framework of reinforcement learning includes environment, agent, state, action, and reward.

Environment: environment by B_i^t, U_j^t and CS. Where, U_j^t is distributed in the ocean and has its own velocity. Each reinforcement learning step corresponds to a time slot in the environment, B_i^t sends its extracted state from the environment to the agent. The action of agent output will be applied to the network before the next time slot starts so that the state in the environment will change accordingly.

Agent: the state input of agent is the signal-to-noise ratio from the whole network, and the output is (Γ_{thr}, t_{thr}). During the training process, the agent interacts with the environment and updates the weights of the neural network according to the feedback of the reward function.

Status: we set the SNR between U_j^t and B_i^t as the state s_t. SNR not only reflects the network environment but also reflects the motion information of all ships.

Actions: define actions as vectors $\{\Gamma_{thr\_1}, \ldots, \Gamma_{thr\_K}, t_{thr\_1}, \ldots, t_{thr\_X}\}$. Where Γ_{thr_k}, $k = 1, \ldots, K$ represents the possible value of the SNR threshold used to calculate the offload policy. t_{thr_x}, $x = 1, \ldots, X$ represents the possible value of the maximum tolerable delay threshold for the offload task. According to the input state, the agent will get the probability of each possible value.

Reward: reward function r_t is closely related to the goal of minimizing the offloading cost, $r_t = -\sum_{t=1}^{T} \sum_{j=1}^{M(t)} Cost_j(t)$

For example, in Algorithm 2, this paper uses actor-critic algorithm to train agents that can generate dynamic priority for maritime offloading. Actor in actor-critic algorithm is a policy network $V_\varphi^\pi(s_t)$, the probability of each action can be given according to the state, Critic is a state value network $V_\varphi^\pi(s_t)$, according to the status, the predicted value of reward can be given. The basic architecture of actor network and critic network is similar, and the difference between them lies in the output layer. The input state first passes through the one-dimensional convolution layer and then flattens and inputs to the fully connected layer. According to the defined action space, the dimension of the actor output layer is $(K + X)$, where the output layer of $1 \sim K$ dimension represents each probability of Γ_{thr}, the output layer of $(K + 1) \sim (K + X)$ dimension characterizes each probability of t_{thr}. The dimension of critic output layer is 1, and only the reward value is output. In the process of training, actor network interacts with the environment to gain experience (s_t, a_t, r_t, s_{t+1}). When the accumulated experience reaches the maximum number of batches or time slots in the environment, the neural network weights will be updated. After the training, the agent can output the best system parameters in the way of ε-greedy to realize the maritime computation offloading based on dynamic priority.

Algorithm 2 Dynamic priority based on actor-critic

1) Initialization: The weight of $\pi_\theta(s_t)$ and $V_\varphi^\pi(s_t)$, Superparameter of neural network

2) while the convergence condition is not completed && the maximum number of iterations is not reached.

3)   Clear experiences

4)   while do

5)   Input the global signal-to-noise ratio of the whole network into $\pi_\theta(s_t)$

6)   Output Γ_{thr}, t_{thr} by using the ε-greedy method

7)   According to $\Gamma_{thr}, t_{thr}, r_t$ is obtained by calculating the offloading cost function

8)   Add r_t to experiences

9)   Get cumulative discount reward value $\sum \gamma^{t'-t} r_{t'}$ by $V_\varphi^\pi(s_t)$

10)   if the length of experience base reaches the number of batches, or the time slice counter in the environment reaches the maximum value

11)   update the weight of $V_\varphi^\pi(s_t)$

12)   update the weight of $\pi_\theta(s_t)$

13)   end if

7)   end while

8) end while

---

4 Simulation Experiment

In this section, we first introduce the simulation experiment settings, then observe and analyze the performance of the proposed algorithm and the comparison algorithm in the offloading delay, user energy consumption, and the response rate of the offloading task from the experimental results, and finally analyze the convergence of the proposed algorithm.

4.1 Experimental Setting

Suppose that the number of EN ships $N(t) = 20$, the number of UE $M(t) = 60$, the total number of time slots $t = 60$, and the length of each time slot t is 1 s. The offloading task requests of UE follow the Poisson process, and the arrival rate is 12 per minute. Offload task data size $b(\theta_j(t))$ obeys the uniform distribution of [300, 800] kb. The computing resource $d(\theta_j(t))$ required to offload the task obeys the uniform distribution of [300, 800] megacycles, and the maximum tolerable delay of the offloading task is $delay^{max}(\theta_j(t))$ obeys the uniform distribution of [1, 5] s. The bandwidth of the shipborne base station is 20 MHz, and the R_{max} of each shipborne base station is 100. The height of UE h_j is 2 m, and the height of the Shipborne base station h_i is 5.5 m. U_j^t transmitting power $Pt(j, i, t)$ is 500 mW, the base station transmitting power $Pt^b(i, j, t)$ is 20 W, and the energy consumption per unit task cycle K_j is 240 μ J. In the path loss formula, the wavelength λ is 0.016 m, and the noise power $\sigma = -100$ dBm. The satellite transmission delay rtt is 480 ms. The default computing resource f assigned to the offloading task by the shipboard edge node $f_{i,j}^e(t)$ is 3.2×10^3 megacycles, and the computing resource capacity F_{max} of shipborne edge nodes is 3.2×10^4 megacycles. Computing resources allocated to each user by terrestrial cloud server $f_j^c(t)$ is 1×10^5 megacycles, assuming that the computing resources of the terrestrial cloud server are infinite.

The number of filters in the one-dimensional convolutional layer of the neural network is 128, and the number of neurons in the fully connected layer is 256. The action space includes 6 Γ_{thr} values and 3 t_{thr} values, so the output layer dimension of the Actor network is 9. The remaining parameters are: the discount coefficient is 0.9, the optimizer is RMSProp, the batch number is 32, the Actor network learning rate is 0.00001, and the Critic network learning rate is 0.00001. The set of Γ_{thr} is {−6, −2, 2, 6, 8, 12}, the set of t_{thr} is {2, 3, 4}. This paper collects 1000 pieces of the real ship sailing data in 60 time slots as training data, another 10 pieces of data as a verification set, and another 100 pieces of data as a test set. The ship data includes the voyages of 20 EN ships (tankers, cargo ships, and work ships are regarded as EN ships) and 60 user ships.

In the above-mentioned experimental environment, we compare the Offloading Algorithm for Maritime Computing (DPCO) based on dynamic priority in this paper with two baseline algorithms in terms of offloading delay, user energy consumption, and offloading task response rate. The two Baseline algorithm ideas under the integrated satellite-maritime mobile edge computing framework are briefly described as follows: (1) MCC algorithm, all computing tasks are offloaded to the terrestrial cloud server, (2) NoRA algorithm, cloud-edge collaboration without resource allocation, all offloaded computing tasks are allocated a default amount of edge computing resources. If the edge computing resources are exhausted, the tasks will be offloaded to the terrestrial cloud server.

4.2 System Offloading Performance

Figure 2 shows the performance of the three algorithms on three important system offloading indicators. Among them, Fig. 2(a) is the offloading delay, Fig. 2(b) is the user energy consumption, and Fig. 2(c) is the offloading task response rate. It can be seen that the offloading delay of the MCC algorithm, user energy consumption is high, and the response rate of offloading tasks is very low. This is because the MCC algorithm first offloads all the computing tasks that can be offloaded to the shipborne edge node and then uploads to the terrestrial cloud server. In this process, many satellite link delays and user communication energy consumption will be generated, and the high delay makes all Offloading tasks difficult to complete within the maximum tolerable delay, so the response rate of offloading tasks is very low. Combined with edge computing, the NoRA

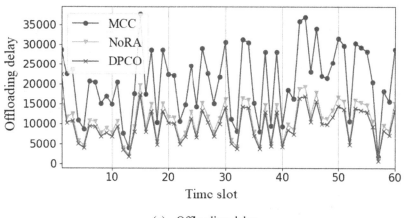

(a)   Offloading delay

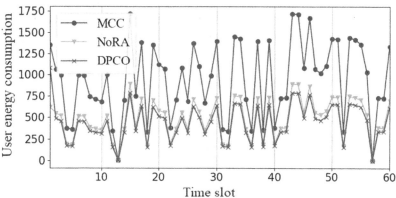

(b)   User terminal energy consumption

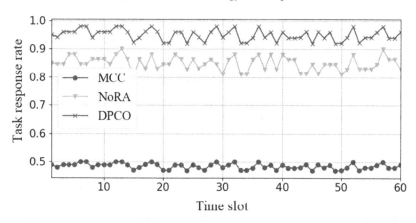

(c)   Offloading task response rate

Fig. 2. Several important system offloading performance.

algorithm's offload delay, user energy consumption, and offloading task response rate is significantly better than the MCC algorithm. The algorithm in this paper outperforms the NoRA algorithm in three offloading indicators. In NoRA, urgent tasks cannot get more computing resources, resulting in tasks that cannot be completed quickly. This paper's algorithm makes full use of the computing resources of terrestrial cloud servers and shipborne edge nodes. It allocates more computing resources to urgent tasks based on the dynamic offloading priority generated by reinforcement learning, which improves offloading performance.

4.3 Algorithm Convergence

Figure 3 shows the convergence curve of the neural network algorithm used to generate the dynamic offloading priority during the training process. We use the number of experiences as the x-axis and the standardized average reward on the verification data as the y-axis, showing the convergence curve of M(t) = 20, 40, 60. It can be seen that the convergence speed of the algorithm is related to the size of the state space (that is, the number of users). The more the number of users, the slower the convergence speed of the algorithm. The three curves can remain basically unchanged in the end. Therefore, the maritime computation offloading algorithm based on dynamic priority in this paper can converge within the limited experience.

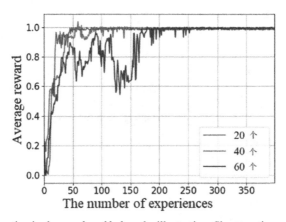

Fig. 3. A figure caption is always placed below the illustration. Short captions are centered, while long ones are justified. The macro button chooses the correct format automatically.

5 Conclusion

This paper mainly studies the offloading method of maritime mobile edge computing, proposes an integrated satellite-maritime mobile edge computing framework, establishes a maritime mobile edge computing model, and proposes a maritime computation offloading method based on dynamic priority. This method uses reinforcement learning to dynamically output a threshold that characterizes the urgency of offloading tasks, and offloads tasks based on this dynamic threshold and intelligently allocates the computing resources of terrestrial cloud servers and shipboard edge servers. Experimental results show that the proposed algorithm can effectively reduce the energy consumption and delay of the maritime mobile edge computing network while maintaining a high task response rate. In future work, we will consider the joint allocation of satellite communication resources, shipborne communication resources, and computing resources under this architecture to promote the green energy-saving of the integrated satellite-maritime mobile edge network and further improve its cloud-edge offloading efficiency.

References

1. Lytra, I., Vidal, M., Orlandi, F., et al.: A big data architecture for managing oceans of data and maritime applications. In: 2017 International Conference on Engineering, Technology and Innovation (ICE/ITMC), pp. 1216–1226 (2017)
2. Feng, W., Tang, R., Ning, G.E.: Perspectives on coordinated satellite-terrestrial intelligent maritime communication networks. Telecommun. Sci., 1–15 (2020)
3. Kim, Y., Song, Y., Lim, S., et al.: Hierarchical maritime radio networks for internet of maritime things. IEEE Access **7**, 2169–3536 (2019)
4. Alwarafy, A., Al-Thelaya, K., Abdallah, M.: A survey on security and privacy issues in edge computing-assisted Internet of Things. IEEE IOT J., 1–1 (2020)
5. Li, J., Gao, H., LV T, et al. Deep reinforcement learning based computation offloading and resource allocation for MEC. In: 2018 IEEE Wireless Communications and Networking Conference (WCNC), pp. 1–6 (2018)
6. Wang, J., Hu, J., Min, G., et al.: Computation offloading in multi-access edge computing using a deep sequential model based on reinforcement learning. IEEE Commun. Mag. **57**, 64–69 (2019)
7. Du, J., Zhao, L., Feng, J., et al.: Computation offloading and resource allocation in mixed fog/cloud computing systems with min-max fairness guarantee. IEEE Trans. Commun. **66**, 1594–1608 (2018)
8. Huang, M., Liu, W., Wang, T., et al.: A cloud-MEC collaborative task offloading scheme with service orchestration. IEEE IOT J. **7**, 5792–5805 (2019)
9. Munasinghe, K.S., et al.: Traffic offloading 3-tiered SDN architecture for dense nets. IEEE Netw. **31**, 56–62 (2017)
10. Hosseinzadeh, M., Tho, Q., Ali, S., et al.: A hybrid service selection and composition model for cloud-edge computing in the Internet of Things. IEEE Access **8**, 85939–85949 (2020)
11. Xiao, A., Ge, N., Yin, L., et al.: Adaptive shipborne base station sleeping control for dynamic broadband maritime communications. In: 2017 19th Asia-Pacific Network Operations and Management Symposium (APNOMS), pp. 7–12 (2017)
12. Teixeira, F.B., et al.: Enabling broadband internet access maritime using tethered balloons: the BLUECOM+ experience. IEEE OCEANS **2017**, 1–7 (2017)
13. Jo, S.W., Shim, W.S.: LTE-maritime: high-speed maritime wireless communication based on LTE technology. IEEE Access **7**, 53172–53181 (2019)

Research on Invulnerability of Spatial Information Networks Based on Improved Jump-Range-Node Method

Shuang Hu[1], Ming Zhuo[2], Peng Yang[1(✉)], Simin Wan[1], and Zhiwen Tian[2]

[1] School of Mathematical Sciences, University of Electronic Science and Technology of China,
Chengdu 611731, China
`Yang_peng_26@163.com`
[2] School of Information and Software Engineering,
University of Electronic Science and Technology of China, Chengdu 611731, China

Abstract. With the rapid development of spatial information networks, its applications become more and more popular, and the requirement of the invulnerability of the spatial information network is also improved. Therefore, how to measure the invulnerability of spatial information networks efficiently and accurately has become a hot research topic for current scholars. Firstly, we introduce the research status of invulnerability of spatial information networks. Secondly, considering that the high dynamic characteristics of the spatial information network topology, on the basis of the research on the survivability of complex networks, a new survivability evaluation method for spatial information networks was proposed. The improved jump-range-node method introduces the index of closeness centrality and redefines the normalization factor. The simulation results show that the improved jump-range-node method can effectively distinguish the importance of spatial information network nodes with similar topological structures, and at the same time can reasonably evaluate the invulnerability of spatial information networks.

Keywords: Spatial information network · Invulnerability · Node importance · Jump-range-node method · Closeness centrality

1 Introduction

Space information network is a global information network formed by high, medium and low orbit satellites, stratospheric balloons, manned or unmanned aerial vehicles and other equipment, which can complete the acquisition, transmission and processing of information on the ground, in the air and in deep space [1–4]. As a national infrastructure, spatial information network has many outstanding advantages, such as wide coverage, flexible networking and wide applications. It plays an important role in the service of ocean navigation, emergency rescue, navigation and positioning, aerospace measurement and control, etc. It also expands human science, culture and production activities to the ocean, deep space and many other spaces.

© Springer Nature Singapore Pte Ltd. 2021
Q. Yu (Ed.): SINC 2020, CCIS 1353, pp. 84–93, 2021.
https://doi.org/10.1007/978-981-16-1967-0_6

Taking into account the exposure of the spatial information network, its complex network structure and higher networking technology requirements. In addition, its maintenance is far more difficult than terrestrial networks. Once a node fails or is maliciously attacked, the whole networks may face the risk of paralysis. Therefore, the network survivability assessment is of great significance to the construction of the network.

Network survivability refers to the ability of the network to maintain or restore its performance to an acceptable level when some nodes or links in the network are deliberately attacked and cause some nodes or links to fail. It is an important attribute to measure network topology [5–8]. Network invulnerability can be obtained by evaluating the importance of network nodes. After accurately evaluating the importance of network nodes, the invulnerability of the whole network can be improved by protecting these key nodes.

2 Related Work

2.1 Overview of Network Invulnerability

In recent years, the research on network invulnerability has emerged in an endless stream. For example, the invulnerability of the network is measured by the connectivity [9], integrity [10], adhesion [11] and expansion coefficient [12]. However, these invulnerability evaluation indicators are based on a global perspective, ignoring the correlation between nodes and links within the network, and the calculation methods are simple, which have certain limitations. With the development of research, more and more researches begin to focus on the relationships between nodes and links in the network. Wu et al. [13] calculated the weighted sum of different closed loop numbers in the network from the degree of nodes, and proposed the concept of natural connectivity degree. Wu et al. [14] decomposed the natural connectivity of the network into local natural connectivity and quantified their contribution to the robustness of the network. Some scholars have studied the invulnerability of Walker Constellation based on natural connectivity and probability matrix [15, 16]. Liu [17] proposed a network invulnerability method based on the number of shortest path number from node links. Combined with the concept of hop-surface, Guo [18] proposed jump-range-node method. Similarly, only from the local will ignore the global nature of the networks, so a single invulnerability evaluation index has been unable to meet the needs of dynamic changes in the networks. Therefore, many researches combined the global information with the local information and proposed a new composite invulnerability evaluation index.

2.2 Jump-Range-Node Method

According to the invulnerability of tactical communication network, Guo [18] proposed jump-range-node method. Given an undirected network $G = (V, E)$, $V = \{v_1, v_2, \cdots, v_n\}$ which represents the set of all nodes; $E = \{e_1, e_2, \cdots, e_n\}$ stands for the set of nodes' connection edges. A surface consisting of all nodes with the same hop number is called hop-surface. All nodes with the same distance from a node i are called hop-surface nodes with the same hop number. The principle of jump-range-node method is as follows. Divide the hop-surface according to the distance from a certain node to other nodes, and then define the sum of the reliability of the node to all hop-surfaces as the importance of the node. Finally, the mean value of the importance of all nodes is defined as the invulnerability of the whole network.

In order to simplify the complexity of evaluation, the jump-range-node method neglects the connection of the nodes in the hop-surface. As shown in Fig. 1, in calculating the importance of node 1, this method only considers the number of links between node 1 and the 1-th hop-surface and the number of links between the 1-th hop-surface and the 2-th hop-surface, but does not consider the internal links of the 1-th hop-surface, which is obviously inaccurate.

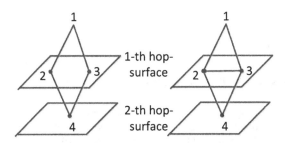

Fig. 1. Node 1 hop-surface model

Then, Wang et al. [19] improved the jump-range-node method and considered the influence of the hop-surface internal links on the invulnerability of the network. However, this method still neglects the link connection structure between the hops, which makes it difficult to distinguish the nodes when evaluating their importance. As shown in Fig. 2, according to Ref. [19], the importance of node 1 in both networks is 0.818. This is because not all intra-hop links have an effect on survivability.

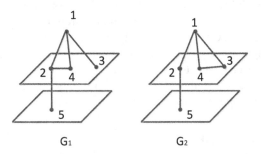

Fig. 2. Node 1 hop-surface model with different topological structure

Wei et al. [20] improved the problem of indistinguishability of similar topological structures based on the above method. The improved jump-range-node method not only considers the influence of the number of links between two hop-surfaces and the connection situation on the network reliability, but also considers the influence of the number of links within the hop-surface and its distribution structure on the network reliability, which effectively solves the problem that the importance degree of nodes cannot be distinguished when evaluating the importance of nodes.

3 Our Method

3.1 Improvement Ideas

Ref. [20] improves the effectiveness of the evaluation method by dividing the links, but the results are not accurate when calculating the node importance of complex spatial information networks. Calculate the node importance of six nodes in Fig. 3 and the results are shown in Table 1, in which the numbers in brackets indicate the order of node importance.

As far as the ranking method is concerned, if the frequency of nodes with the same ranking is lower, the method is more effective. Furthermore, for spatial information networks, it is difficult to prove which node or nodes are of the greatest importance if not for specific purposes. Therefore, this paper adopts the method of "voting" to determine the ranking result of node importance in the network. The voting rules are as follows: If more than half of the algorithms consider node A to be more important than node B, and then take this result as the importance ranking result.

We can see that the node importance ranking results of Ref. [20] are different from closeness centrality and Ref. [18]. The experimental results of closeness centrality and

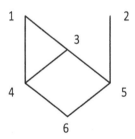

Fig. 3. Network topology

Table 1. The node importance evaluation results in Fig. 3

Node	Closeness centricity	Ref. [18]	Ref. [20]
1	0.556 (5)	0.488 (5)	0.508 (5)
2	0.455 (6)	0.304 (6)	0.304 (6)
3	0.714 (1)	0.720 (1)	0.728 (1)
4	0.625 (3)	0.696 (3)	0.702 (2)
5	0.714 (1)	0.720 (1)	0.680 (3)
6	0.625 (3)	0.560 (4)	0.560 (4)

Ref. [18] show that node 5 is more important than node 4, while Ref. [20] believes that node 4 is more important. There are two reasons for this result: On the one hand, when calculating the reliability between a node and other hops, the links between nodes in the same hops will be included, which results in the superposition of computation. On the other hand, according to the definition of jump-range-node method, the normalized factor is mainly determined by the link distribution between two adjacent hop-surfaces. It only considers the local information of the network and ignores the influence of each node in the network on the reliability of the global network topology, which leads to inaccurate evaluation results. In our paper, the normalized factor is improved, and the closeness centrality is introduced into the normalized factor.

Closeness centrality is a global metrics, which reflects the closeness of a node to other nodes. In other words, if one node is closer to other nodes, its closeness centrality will be greater, and then the node will be more important. Besides, closeness centrality reflects the relationship between nodes, which makes up for the shortcoming of jump-range-node method, which only considers the relationship between links without considering the relationship between nodes, and improves the accuracy of measurement results. The improved normalized factor combines the local information with the global information, which makes the measurement result more accurate.

3.2 Algorithm

On the basis of Ref. [20], this paper combines jump-range-node method with the complex network theory, and considers introducing closeness centrality into the normalized factor. Closeness centrality is defined as follows.

Definition 1. d_i is the average of the distances between node i and other nodes, defined as

$$d_i = \frac{1}{n-1} \sum_{j=1}^{n-1} d_{ij} \tag{1}$$

Closeness centrality (CC) of node i is meant by the reciprocal of the sum of d_i, defined as:

$$CC_i = \frac{1}{d_i} = \frac{n-1}{\sum_{j=1}^{n-1} d_{ij}} \tag{2}$$

Where d_{ij} denotes the shortest path length between node i and node j. If node i is not connected to node j, then $d_{ij} = \infty$.

The proposed method not only combines the global information with the local information, but also synthetically evaluates the invulnerability of the network from different perspectives, which improves the accuracy of the evaluation. The algorithm is as follows:

Definition 2. The invulnerability R_G of the network $G(N, E)$ is the sum of the average reliability of each node to all other hops, defined as:

$$R_G = \frac{1}{N} \sum_{I=1}^{N} r_i \, , \, r_i = \sum_{j=1}^{M} r_{ij} \tag{3}$$

Where r_i is the sum of the reliability of node i to other surface-hops. M represents the maximum number of hops from node i. r_{ij} is the reliability between node i and the j-th hop-surface. The formula is as follows:

$$\begin{cases} r_{i1} = r_{01} \\ r_{i2} = r_{i1} \bullet r_{12} \\ r_{i3} = r_{i2} \bullet r_{23} \\ \cdots \\ r_{i(m+1)} = r_{im} \bullet r_{m(m+1)} \end{cases} \tag{4}$$

Where $r_{m(m+1)}$ is the normalized reliability from the m-th hop-surface to the $(m + 1)$-th hop-surface. We assume that nodes and links are reliable, and then $r_{m(m+1)} = \mu_{ij} r_{l_m} = \mu_{ij}$. Thus, knowing the network topology, we can calculate the invulnerability of the network.

The normalized factor μ'_{ij} of the jump-range-node method is:

$$\mu'_{ij} = \frac{|V_{i(j+1)}|}{|V| - 1} \bullet \frac{|L_{ij}|}{|V_{ij}| \bullet |V_{i(j+1)}|} \tag{5}$$

Then, considering the effects of intra-hopper links, the number of node connections between hop-surfaces and the structure of intra-hopper links on the reliability of the lower hop-surfaces, the normalized factor is modified as:

$$\mu''_{ij} = \frac{|L_{ij}|}{|V_{ij}|(|V| - 1)} \bullet \frac{|Q_{ij}|}{|V_{ij}|} \bullet (1 + \frac{|T_{ij}| \big/ C^2_{|V_{ij}|}}{|V| - 1}) \tag{6}$$

Furthermore, we consider introducing closeness centrality into the normalized factor. We combined the global and local information to evaluate the importance of nodes from the perspectives of nodes and links, and redefined the normalized factors as follows:

$$\mu_{ij} = \frac{|L_{ij}|}{|V_{ij}|(|V| - 1)} \bullet \frac{|Q_{ij}|}{|V_{ij}|} \bullet (1 + \frac{|T_{ij}| \big/ C^2_{|V_{ij}|}}{|V| - 1}) \bullet CC(i) \tag{7}$$

Where V_{ij} denotes the set of all nodes starting at node i with hops of j; L_{ij} represents the set of links connecting V_{ij} and $V_{i(j+1)}$. P_{ij} is the set of nodes in V_{ij} that have edges connected to the nodes in $V_{i(j+1)}$. P'_{ij} is the set of nodes in V_{ij}, which is not connected to the $(j + 1)$-th hop-surface, but has path with nodes in the set P_{ij}, $Q_{ij} = P_{ij} \cup P'_{ij}$. T_{ij} is the set of links between nodes in the set Q_{ij}.

4 Simulation and Analysis

The topology of spatial information network is complex, and the network is in constant change and movement. It is difficult to use mathematical analysis and numerical analysis methods to evaluate the invulnerability of spatial information network, so we adopt the method of simulation to test and evaluate the performance of the network. Firstly, we use STK (Satellite Tool Kit) to construct the spatial information network system, and then use Matlab for numerical calculation. The parameters of the spatial information network system are shown in Table 2:

Table 2. Spatial information network parameters

Satellite number	Altitude(km)	Inclination(°)	Trueanomaly(°)	RAAN(°)
1	6678.14	28.5	310	120
2	6678.14	28.5	240	220
3	10000.00	0.0	0	0
4	10000.00	0.0	90	0
5	10000.00	0.0	180	0
6	10000.00	0.0	270	0

We assume that any two satellites within the line of sight in the space information network are linked. Because of the high dynamic characteristic of spatial information network, the network topology structure is changing at any time. Many of these topologies are very similar, so how to distinguish the networks with similar topologies effectively becomes the key to evaluate the invulnerability method. Therefore, we select three similar topologies, as shown in Fig. 4:

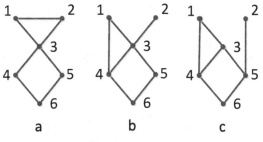

Fig. 4. Topology

4.1 Node Importance Experiment

Firstly, we calculate the node importance of Fig. 4(b) using the method proposed in this paper and compare the results with those of Ref. [18], Ref. [20] and closeness centrality. The results are shown in Table 3:

Table 3. Figure 4(b) Results of nodes importance assessment

Node	Closeness centrality	Ref. [18]	Ref. [20]	Our method
1	0.625 (3)	0.520 (5)	0.544 (4)	0.306 (4)
2	0.500 (6)	0.336 (6)	0.337 (6)	0.132 (6)
3	0.833 (1)	0.880 (1)	0.862 (1)	0.710 (1)
4	0.714 (2)	0.720 (2)	0.728 (2)	0.494 (2)
5	0.625 (3)	0.560 (3)	0.560 (3)	0.312 (3)
6	0.556 (5)	0.532 (4)	0.534 (5)	0.262 (5)

As shown in Table 3, the four methods have the same sorting results for node 3, node 4, and node 5. Node 3 is considered to be the most important, followed by node 4 and node 5. However, there are different results in the evaluation of node 1. Closeness centrality, Ref. [20] and our method believe that node 1 is more important than node 6, contrary to Ref. [18]. According to the voting rules, we can see that the accuracy of our method is better than that of Ref. [18]. In addition, closeness centrality cannot effectively distinguish the importance of node 1 and node 5. In contrast, our method can effectively distinguish the importance of all nodes, and the result of ranking is consistent with the result of comprehensive comparison. It is proved that our method is superior to other methods in accuracy and validity.

Further, we calculated the importance of node 1 in Fig. 2 and compared it with the closeness centrality and Ref. [18]. The calculated results are shown in Table 4.

Table 4. Comparison of the importance of node 1 in G_1 and G_2

Node	Closeness centrality	Ref. [18]	Ref. [19]	Our method
Node 1 in G_1	0.800	0.813	0.818	0.629
Node 1 in G_2	0.800	0.813	0.818	0.613
Distinction	No	No	No	Yes

By analyzing the data in Table 4, we can see that the proposed method can effectively distinguish the importance of nodes in the network with similar topological structure, and it is suitable for the spatial information network with high dynamic.

4.2 Network Invulnerability Experiment

In this experiment, we evaluated the network invulnerability of the three network topologies with similar topological structure by using the methods of Ref. [18], Ref. [19], Ref. [20] and closeness centrality, and the results are shown in Table 5.

Table 5. Invulnerability results

Model	Ref. [18]	Ref. [19]	Ref. [20]	Closeness centrality	Our method
Figure 4(a)	0.5840	0.5908	0.5645	0.6157	0.3423
Figure 4(b)	0.5913	0.5977	0.5974	0.6422	0.3694
Figure 4(c)	0.5813	0.5870	0.5794	0.6148	0.3420

As can be seen from Table 5, when evaluating the network invulnerability, the results of the proposed and the Ref. [18], Ref. [19] and closeness centrality all agree that the network invulnerability of Fig. 4(b) is the strongest, followed by that of Fig. 4 (a), Fig. 4(c) is the weakest. However, Ref. [20] considers that Fig. 4(a) has the worst invulnerability, which is not consistent with the comprehensive results.

The proposed method can not only get more accurate ranking results, but also reduce the frequency of the same ranking nodes. In addition, it can distinguish the importance of network nodes with similar topological structure. When the topology structure of high dynamic spatial information network changes slightly, we can also find its rules, so that we can take corresponding measures to implement node protection and improve the network's invulnerability. Besides, the proposed method is more accurate and effective when evaluating the whole network invulnerability.

5 Conclusion

In view of the high dynamic characteristics of spatial information networks, this paper proposes a new method for evaluating network invulnerability. This method introduces the closeness centrality on the basis of the research of the jump-range-node method, and integrates the evaluation index of invulnerability from different perspectives, and obtains the reasonable result. The results of simulation experiments show that our method can not only improve the accuracy of the ranking results, but also reduce the frequency of the same sorting nodes, which provides a theoretical basis for protecting the key nodes in the spatial information network. Besides, the experimental results also show that our method is more reasonable and accurate for the evaluation of the whole invulnerability of spatial information networks.

References

1. Zhang, T., et al.: Application of time-varying graph theory over the space information networks. IEEE Netw. **34**, 179–185 (2020)

2. Ding, G., Li, L., Wang, J., et al.: Tutorial on big spectrum data analytics for space information networks. J. Wireless Commun. Netw. **2018**, 262 (2018)
3. Xu, D., et al.: Spatial information theory of sensor array and its application in performance evaluation. Commun. lett. **13**, 2304–2312 (2019)
4. Ma, J.L.: Research on Optimization and Invulnerability of Spatial Information Network. Xidian University, Xi'an (2018)
5. Zhang, J.N., et al.: A new method of evaluation of LEO satellite communication network survivability. Comput. Digital Eng. **42**(9), 1645–1648 (2014)
6. He, X.: Study on Satellite Communication System's Reliability Based on Invulnerability. University of Electronic Science and Technology, Chengdu (2013)
7. Wang, Y.P.: Research on Survivability of Satellite Network Topology. Dalian University of Technology, Dalian (2018)
8. Wei, D.B., et al.: The important node assessment method of satellite network based on near the center. In: 2016 International Conference on Network and Information Systems for Computers (ICNISC), pp. 103–107 (2016)
9. Frank, H., et al.: Analysis and design of survivable networks. IEEE Trans. Commun. Technol. **18**, 501–519 (1970)
10. Wilkov, R.: Analysis and design of reliable computer networks. IEEE Trans. Commun. **20**, 660–678 (1972)
11. Boesch, F., et al.: On graphs of invulnerable communication nets. IEEE Trans. Circ. Theory **17**, 183–192 (1970)
12. Ruan, Y.R., et al.: Node importance measurement based on neighborhood similarity in complex network. Acta Physica Sinica, 371–379 (2017)
13. Wu, J., et al.: Analysis of invulnerability in complex networks based on natural connectivity. Complex Syst. Complex. Sci. 183–192 (2014)
14. Wu, J., et al.: Robustness of regular ring lattices based on natural connectivity. Int. J. Syst. Sci. **42**(7), 1085–1092 (2010)
15. Zhuo, M., et al.: A method for evaluating the invulnerability of space-sky information network. In: 2019 IEEE 2nd International Conference on Automation, Electronics and Electrical Engineering (AUTEEE), pp. 412–417 (2019)
16. Wan, S.M., et al.: Survivability simulation of walker constellation based on probability matrix. In: 2019 Software-Defined Satellite symposium (2019)
17. Liu, H.B.: Invulnerability analysis of complex ship network based on the shortest path. Ship Sci. Technol. 163–165 (2018)
18. Guo, W.: Reliability evaluation method of tactical communication network. Acta Electronica Sinica **26**(1), 3–6 (2000)
19. Wang, L., et al.: Evaluation of regional communication network in invulnerability based on improved jump-range-node method. Mod. Electron. Tech. 13–15 (2013)
20. Wei, D.B., et al.: Research on satellite network topologies survivability evaluation method. Comput. Sci. 301–303 (2016)

LoRa Differential Modulation for LEO Satellite IoT

Chengyang Liu⬤, Tao Hong$^{(\boxtimes)}$, Xiaojin Ding, and Gengxin Zhang

Nanjing University of Posts and Telecommunications, Nanjing 210003, China
{1219012322,hongt}@njupt.edu.cn

Abstract. Due to the wide coverage nature of satellite communication system, Satellite-based Internet of Things (S-IoT) has become a new hot topic in the field of IoT. This paper proposed a differential modulation scheme for LoRa signal to over-come the doppler frequency shift (DFS) in satellite-based IoT from point view of modulation technique. Furthermore, a maximum likelihood sequence detection (MLSD) demodulation algorithm is also designed at the satellite receiver to optimize error spread phenomenon of proposed differential modulation scheme. Simulation results show that the proposed differential modulation LoRa signal has better bit-error-ratio (BER) performance compared with traditional LoRa modulation signal in LEO S-IoT.

Keywords: IoT · Satellite IoT · LoRa · Differential modulation · Maximum likelihood sequence detection

1 Introduction

In recent years, the Internet of Things (IoT) applications have developed rapidly in many areas, such as smart buildings, personal healthcare, environmental monitoring, logistics transportation and so on. LoRaWAN protocol and standard is widely used in the terrestrial basestation-based IoT at the unlicensed band due to its advantages of low power consumption and wide coverage [1]. However, terrestrial basestation-based IoT services are usually limited by the geographical environment and the natural disaster, such as ocean, remote areas, and earthquake disaster, because terrestrial basestation is difficult to build in these areas or be destroyed by natural disaster. To address this problem, scholars began to research the application potential of LoRaWAN in the LEO satellite-based IoT [2].

Wu focused on the LEO satellite constellation-based IoT services for their irreplaceable functions, and provided an overview of the architecture of the LEO

Supported by the National Science Foundation of China (No. 91738201 and 61971440), the Jiangsu Province Basic Research Project (No. BK20192002), the China Postdoctoral Science Foundation (No. 2018M632347), and the Natural Science Research of Higher Education Institutions of Jiangsu Province (No. 18KJB510030).

Q. Yu (Ed.): SINC 2020, CCIS 1353, pp. 94–105, 2021.
https://doi.org/10.1007/978-981-16-1967-0_7

satellite constellation-based IoT in [3]. Qian presented a symmetric chirp signal with better crosscorrelation performance to overcome high dynamic characteristic in LEO satellite-based IoT in [4]. However, the proposed signal waveform degraded the performance of multiple access. On this basis, Qian proposed an improved asymmetric signal to address the problem of multiple access in [5]. A chirp spread spectrum ALOHA protocol in the MAC layer is also designed in this paper. In [6], Yang proposed a folded chirp frequency shift keying (FCrSK) modulation with strong immunity to doppler frequency shift (DFS).

The traditional LoRa modulation scheme modulates the communication information into the starting frequency of each chirp symbol [7], which was developed under the scenario of terrestrial basestation-based IoT with quasistatic transmission channel. While, bit-error-ratio (BER) performance of this traditional modulation scheme, only relying on its preamble to estimate and compensate the frequency offset, will be degraded under scenario of the LEO satellite-based IoT with high dynamic characteristic. To solve this problem, we propose a differential modulation scheme for LoRa signal to overcome the doppler frequency shift in satellite-based IoT from point view of modulation compared with the waveform design in [4–6]. The proposed modulation scheme modulates the communication information into the starting frequency relationship of two adjacent symbols compared with the starting frequency of each chirp symbol in traditional LoRa modulation scheme. Furthermore, a maximum likelihood sequence detection (MLSD) demodulation algorithm is also designed at the satellite receiver to optimize error spread phenomenon of proposed differential modulation scheme. Simulation results show that the proposed differential modulation LoRa signal has better BER performance compared with traditional LoRa modulation signal in satellite-based IoT.

The paper is organized as follows. Section 2 describes the LoRa differential modulation and LEO satellite channel model. Section 3 describes demodulation and the MLSD algorithm. At the same time, the process of LoRa frame synchronization and the influence of frequency offset on LoRa modulation are also introduced. Section 4 analyzes the error performance of differential modulation and the performance of the MLSD algorithm. Finally, conclusions are drawn in Sect. 5.

2   Differential Modulation Scheme for LoRa Signal

2.1   System Model

Figure 1(a) shows the scenario diagram of LEO satellite-based IoT. The LEO satellite covers an area with S-IoT terminals, such as desert, ocean and remote area. The IoT terminals transmit short data packets to the LEO satellite via the uplink channel. The satellite receiver demodulates the packets and relay the information to the ground gateway by the inter-satellite link and the feeder link. Figure 1(b) illustrates the flow chart of the transmit signal from the IoT terminal

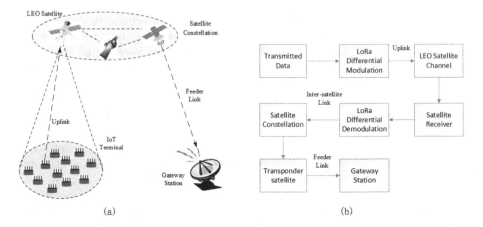

Fig. 1. LEO satellite-based IoT, (a) scenario of LEO satellite-based IoT, (b) flow chart of the transmit signal.

to the ground gateway. Figure 2 shows the satellite coverage model, where e denotes the elevation angle between the terminal and the satellite, s denotes the distance the satellite and terminal, the geocentric angle between the satellite orbital coverage area and the terminal is denoted by α, the angle between the satellite velocity vector and the terminal is denoted by β, R_e denotes the radius of the earth and R_s denotes the distance from the satellite to the center of the earth. In this paper, we focus on designing the modulation scheme in the IoT terminal and the demodulation algorithm in the satellite receiver. Accord to the satellite coverage model, the DFS can be written as:

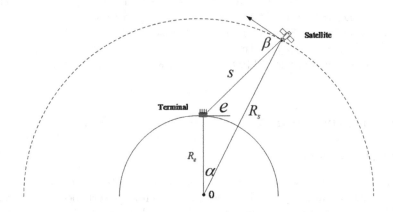

Fig. 2. Satellite coverage model

$$f_D = \frac{f \cdot v \cdot \cos(\beta)}{c} \tag{1}$$

where f is the carrier frequency, v is the relative speed between LEO satellite and the terminal, and c is the speed of light.

2.2 LoRa Differential Modulation

We consider that the transmit information is denoted by $X_n, X_n \in [0, 2^{SF} - 1], n = 0, 1, \cdots, N - 1$, where N is the total number of transmit symbols. Each symbol contains SF bits $(SF = \{7, 8, ..., 12\})$. After the differential operation, the transmit information can be expressed as:

$$\Delta X = \sum_{n=1}^{N} \Delta X_n \tag{2}$$

$$\Delta X_n = (\Delta X_{n-1} + X_n) \bmod 2^{SF}, n = 1, 2, \cdots, N, \Delta X_0 = 0 \tag{3}$$

The starting frequency information of the proposed differential chirp symbol can be written as:

$$f_0 = \Delta X \frac{B}{2^{SF}} + f_{\min} \tag{4}$$

The chirp rate can be denoted as follows:

$$\mu = \frac{B}{T} = \frac{B^2}{2^{SF}} (Hz/s) \tag{5}$$

Therefore, we can obtain the frequency hopping time as:

$$T_0 = \frac{2^{SF} - \Delta X}{B} \tag{6}$$

The proposed differential LoRa signal can be written as:

$$s(t) = \begin{cases} e^{j\pi\mu t^2 + j2\pi f_0 t}, 0 \le t < T_0 \\ e^{j\pi\mu t^2 + j2\pi(f_0 - B)t}, T_0 \le t < T \end{cases} \tag{7}$$

For the sake of convenience, we suppose that the spread factor equals to 7 $(SF = 7)$, the transmit information is $X = [0, 32, 80, 112, 64]$, and differential modulation information can be expressed as $\Delta X = [0, 32, 112, 96, 32]$. Figure 3(b) shows the time-frequency diagram of proposed differential LoRa modulation. Compared with the traditional LoRa modulation signal in Fig. 3(a). We can find that the transmit information is modulated into the relationship between the starting frequency of two adjacent symbols to overcome the large DFS in LEO satellite-based IoT.

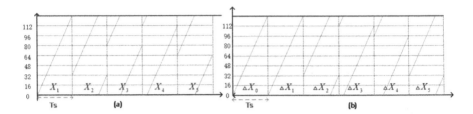

Fig. 3. Time-frequency diagram, (a) traditional LoRa modulation signal, (b) proposed differential LoRa modulation signal.

3   Demodulation Scheme for the Proposed Differential LoRa Signal

3.1   Demodulation Scheme for Traditional LoRa

The demodulation scheme of the traditional LoRa receiver is shown in the Fig. 4. The first step in the receiver is to detect the preamble. The LoRa signal undergoes de-chirp and FFT to obtain a spectrum. The information carried by LoRa symbols can be obtained by searching the peaks of the spectrum. When receiver detects M consecutive symbols have the same modulation information, these symbols are judged as preambles where M denotes the number of the up-chirps in preamble.

The frequency offset caused by the motion of the LEO satellite will result in the peak deviation of the demodulation signal. Therefore, the receiver must synchronize with the receive signal before demodulation the receive signal. According to the influence of frequency offset for demodulation, the frequency offset is divided into two parts: one is integer frequency offset; the other is fractional frequency offset. The integer frequency offset and fractional frequency offset are denoted as follows:

$$\left| \frac{|f_{Dint}|}{f_s} \cdot 2^{SF} \right| = i, i \in Z \tag{8}$$

$$\left| \frac{f_{Dfrac}}{f_s} \cdot 2^{SF} \right| < 1 \tag{9}$$

where f_s denotes the sampling frequency.

The integer DFS can be estimated according to the up-chirp offset in the preamble as:

$$\hat{f}_{Dint} = f_s \cdot \hat{S}_{up}/2^{SF} \tag{10}$$

where $\hat{S}_{up}$ denotes the demodulation information of up-chirp.

The residual fractional DFS must be compensated to prevent the demodulation value of the symbol from being located on two adjacent symbols. The fractional DFS can be estimated as:

$$r[n] = e^{j2\pi n \frac{\hat{S}}{2^{SF}} + j2\pi n \frac{f_{Dfrac}}{f_s}} \tag{11}$$

$$\Delta\phi = 2\pi\frac{f_{Dfrac}}{f_s}.2^{SF} \tag{12}$$

$$\hat{f}_{Dfrac} = f_s \cdot \Delta\phi/2^{SF} \tag{13}$$

If the sync-word is detected after estimating and compensating the frequency offset, the receive achieve the synchronization function for the receive signal.

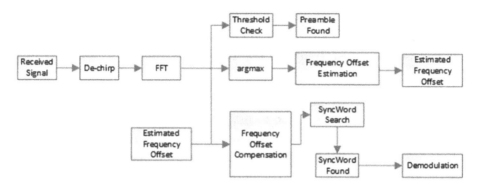

Fig. 4. Receiver scheme of traditional LoRa modulation signal

3.2 Differential Demodulation Scheme for Proposed Signal

The demodulation scheme for proposed differential receiver is shown in the Fig. 5.

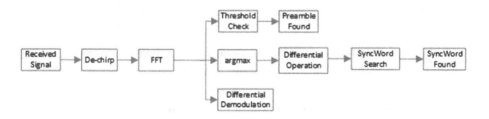

Fig. 5. Receiver scheme of the proposed differential modulation signal

The receive signal can be written as:

$$r(t) = s(t) \cdot e^{j2\pi f_D t} + w(t) \tag{14}$$

where $w(t)$ is AWGN with zero mean and variance N_0. Thus, the digital LoRa signal expression is written as:

$$r[n] = s[n] \cdot e^{j2\pi n\frac{f_D}{f_s}} + w[n] \tag{15}$$

$$f_s = B \tag{16}$$

where $w[k]$ denotes the digital form for AWGN. Then the digital LoRa signal is multiplied by the matched basic down-chirp signal as:

$$g[n] = r[n] \cdot d[n] = e^{j2\pi(\frac{\Delta X}{2^{SF}} + \frac{f_D}{f_s})n} + \tilde{w}[n] \tag{17}$$

$$d[n] = e^{-j\pi \frac{n^2}{2^{SF}} - j2\pi \frac{f_{min}}{B} n}, n = 0, 1, ..., 2^{SF} - 1 \tag{18}$$

where $\tilde{w}[n] = w[n] \cdot d[n]$.

The differential information $\Delta \hat{K}$ can be estimated according to the spectral peak by Discrete Fourier Transform (DFT) as:

$$
\begin{aligned}
g[k] &= \sum_{n=0}^{2^{SF}-1} e^{j2\pi\left(\frac{\Delta X}{2^{SF}} + \frac{f_D}{f_s}\right)n} \cdot e^{-j2\pi \frac{k}{2^{SF}}n} + \tilde{W}[k] \\
&= \sum_{n=0}^{2^{SF}-1} e^{j2\pi\left(\frac{\Delta X - k}{2^{SF}} + \frac{f_D}{f_s}\right)n} + \tilde{W}[k] \\
&= \begin{cases} 2^{SF} + \tilde{W}[k], k = \left(\Delta X + \frac{f_D}{f_s}2^{SF}\right) \bmod 2^{SF} \\ \tilde{W}[k], else \end{cases}
\end{aligned} \tag{19}
$$

$$\Delta \hat{X} = \arg \max_{k}(|g[k]|) \tag{20}$$

Then, the transmit information X_n can be obtain by:

$$\Delta \hat{X} = \left(\Delta X + \frac{f_D}{f_s}2^{SF}\right) \bmod 2^{SF} \tag{21}$$

$$X_n = (\Delta \hat{X}_n - \Delta \hat{X}_{n-1}) \bmod 2^{SF}, n = 1, 2, ...N \tag{22}$$

According to the aforementioned demodulation scheme, the BER performance will be degraded by the error spreading. This phenomenon means that if there is an error in the demodulation, it will cause errors in the subsequent demodulation and causing serious error spreading. In the following subsection, we will design a MLSD scheme to improve the demodulation performance.

3.3   Maximum Likelihood Sequence Detection Scheme for Proposed Signal

The BER of differential demodulation is defined as follows (only considering the continuous error code):

$$P_e' = \sum_{i=1}^{n} (i+1)P_i \tag{23}$$

$$P_i = (1 - P_e)^2 P_e^i \tag{24}$$

where P_e denotes BER of traditional LoRa modulation method under AWGN channel.

$$P_e = 0.5Q\left(1.28\sqrt{SNR2^{SF}} - 1.28\sqrt{SF} + 0.4\right) \tag{25}$$

$$SNR = \frac{E_s}{N_0 2^{SF}} = \frac{E_b}{N_0} \cdot \frac{SF}{2^{SF}} \tag{26}$$

LoRa differential modulation uses the correlation between adjacent symbols and belongs to relative modulation. The symbol-by-symbol demodulation of traditional LoRa modulation does not need to consider the correlation between symbols. The method to solve the problem of error spreading is to use the correlation between symbols after differential modulation. The relative relationship between symbols is demodulated and transferred to the frequency sequence after FFT. Therefore, the MLSD on the frequency sequence results after FFT can effectively solve the problem of error spreading.

The specific processing flow is as follows:

Assuming that, N symbols is transmitted, the send message sequence is $X_1^N = \sum_{n=1}^{N} x_n$. The information sequence obtained after the receiver demodulates the signal $Y_1^N = \sum_{n=1}^{N} Y_n$, where $Y_n = [y_{k1}\ y_{k2}\ \cdots\ y_{kN}]^T$, represents the metric value corresponding to the k-th symbol frequency f_m. The MLSD is to search the demodulated information sequence $\hat{X}_1^N = \sum_{n=1}^{N} \hat{x}_n$ to maximize the conditional probability $P\left(Y_1^N | \hat{X}_1^N\right)$.

Let $\Phi_N = P\left(Y_1^N | \hat{X}_1^N\right)$ denote the conditional probability, the grid state of the LoRa differential system at time k is denoted by S_k, which is the k-th symbol start frequency information. Therefor the send information $\hat{x}_n$ is uniquely determined by (S_{k-1}, S_k):

$$\hat{X}_1^N = \arg\max \Phi_N = \arg\max \prod_{k=1}^{N} P\left(Y_k | S_{k-1}, S_k\right) \tag{27}$$

If $S_{k-1} = s'$, $S_k = s$, the frequency information transmitted at time k is denoted by f_{ki} and for the k-th symbol, the received sequence $Y_k = [y_{k1}\ y_{k2}\ \cdots\ y_{kN}]^T$ is the square of the FFT output amplitude sequence. Therefor $y_{km}(m \neq i)$ obeys the central χ^2 distribution with 2 degrees of freedom and y_{ki} obeys the noncentral χ^2 with 2 degrees of freedom distribution, and they are all statistically independent, which is:

$$P(y_{km}|f_{ki}) = \begin{cases} \frac{1}{2}\exp\left(-\frac{1}{2}\left(y_{ki} + 2\frac{E_s}{N_0}\right)\right) I_0\left(\sqrt{\frac{2E_s}{N_0} y_{ki}}\right), & m = i \\ \frac{1}{2}\exp\left(-\frac{y_{km}}{2}\right) & m \neq i \end{cases} \tag{28}$$

where $I_0(x)$ denotes the first kind of zero-order modified Bessel function:

$$P\left(Y_k | S_{k-1} = s', S_k = s\right) = \prod_{m=0}^{N-1} P\left(y_{km}|f_{ki}\right)$$

$$= \frac{1}{2^N}\exp\left(-\frac{\frac{2E_s}{N_0} + \sum_{m=0}^{N-1} y_{km}}{2}\right) I_0\left(\sqrt{\frac{2E_s}{N_0} y_{ki}}\right) \tag{29}$$

Then the total metric of the path can be expressed as:

$$\Phi_N = P\left(Y_1^N|\hat{X}_1^N\right) = \prod_{k=1}^{N} P\left(Y_k|S_{k-1}, S_k\right) = C\prod_{k=1}^{N} I_0\left(\sqrt{\frac{2E_s}{N_0}}y_{ki}\right) \qquad (30)$$

where C is a fixed value. Thus:

$$\hat{X}_1^N = \arg\max \Phi_N = \arg\max \prod_{k=1}^{N} I_0\left(\sqrt{\frac{2E_s}{N_0}}y_{ki}\right) \qquad (31)$$

This is the better demodulation scheme for LoRa differential signal sequence detection under the maximum likelihood criterion compared with different demodulation scheme in the Subsect. 3.2. Compared with searching all possible sequences, using the Viterbi algorithm to search for surviving paths can greatly simplify the computational complexity.

4   Simulation Results and Discussion

In order to evaluate the performances of the traditional LoRa modulation signal and proposed differential LoRa modulation signal, we consider the simulation parameters as follows: signal bandwidth $B = 125\,\text{KHz}$, $SF = 10$. Figure 6 shows the DFS influence on the receive signal without considering the AWGN signal. It is found that: the integer DFS causes the spectral peak shifting. While the fractional DFS results in the frequency point energy to shift to adjacent frequency points which leads the spectral peak height decreasing. The closer the value of the fractional frequency offset is to 1 the more energy the spectral peak drops. And this phenomenon causes an error in the demodulation decision when the value of the fractional DFS is greater than 0.5. Figure 7 shows the impact of different fractional frequency offsets on the BER performance in the AWGN channel. It is found that the larger the fractional frequency offset, the more serious the

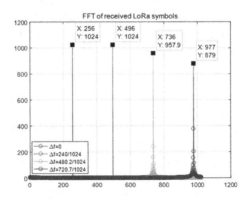

Fig. 6. DFT results by demodulating a symbol $X = 256$ with DFS

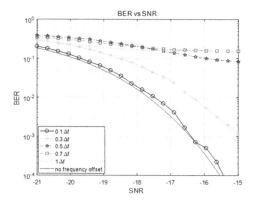

Fig. 7. BER performance of a satellite receiver with fractional DFS

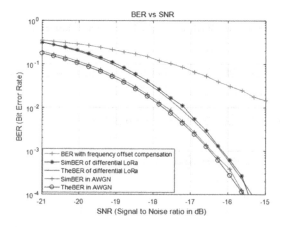

Fig. 8. BER of different modulation (2000 bit)

BER will decrease. Figure 8 shows the BER performance of LoRa traditional modulation in the AWGN channel and the BER performance of LoRa traditional modulation and LoRa differential modulation by different DFS values. The curve SimBER denotes the simulated BER performance by Monte Carlo simulation and curve TheBER denotes the theoretical BER performance. It is found that the traditional modulation scheme cannot accurately estimate the frequency offset which reduces the BER performance. The proposed differential modulation scheme can effectively reduce the impact of DFS. However, the differential demodulation scheme causes the phenomenon of error spreading. Figure 9 shows the BER performance of the MLSD demodulation scheme for the proposed LoRa differential modulation signal. It is found that MLSD algorithm can effectively improve the bit BER performance. This is because the frequency hopping process of the differential LoRa signal has a specific correlation which is equivalent to a encoding in frequency domain. The proposed sequence detection

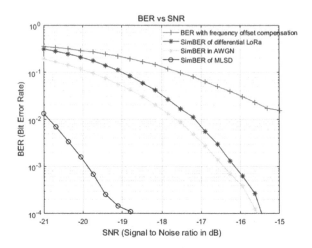

Fig. 9. BER performance of MLSD demodulation scheme

scheme takes advantage of this correlation by correcting a certain percentage of frequency misjudgments to improve BER performance. It can be seen that MLSD demodulation scheme can effectively improve 3 dB BER performance compared with traditional LoRa signal in AWGN channel.

5   Conclusion and Future Work

In this paper, we propose a differential modulation scheme for LoRa signal to solve doppler frequency shift in satellite-based IoT problem. The proposed modulation scheme modulates the communication information into the starting frequency relationship of two adjacent symbols compared with the starting frequency of each chirp symbol in traditional LoRa modulation scheme. Furthermore, a MLSD demodulation algorithm is also designed at the satellite receiver to optimize error spread phenomenon of proposed differential modulation scheme. Simulation results show that the proposed differential modulation LoRa signal has better BER performance.

Interesting topics for future work include massive access methods of the S-IoT and influence of DFS on multi-user uplink.

References

1. Nguyen, T.T., Nguyen, H.H., Barton, R., et al.: Efficient design of chirp spread spectrum modulation for low-power wide-area networks. IEEE Internet Things J. **6**(6), 9503–9515 (2016)
2. Qu, Z., Zhang, G., Xie, J.: LEO satellite constellation for Internet of Things. IEEE Access **5**, 18391–18401 (2017)

3. Wu, T., Qu, D., Zhang, G.: Research on LoRa adaptability in the LEO satellites Internet of Things. In: 2019 15th International Wireless Communications and Mobile Computing Conference (IWCMC) (2019)
4. Qian, Y., Ma, L., Liang, X.: Symmetry chirp spread spectrum modulation used in LEO satellite Internet of Things. IEEE Commun. Lett. **22**(11), 2230–2233 (2018)
5. Qian, Y., Ma, L., Liang, X.: The acquisition method of symmetry chirp signal used in LEO satellite Internet of Things. IEEE Commun. Lett. **23**(9), 1572–1575 (2019)
6. Yang, C., Wang, M., Zheng, L., et al.: Folded chirp-rate shift keying modulation for LEO satellite IoT. IEEE Access **7**, 99451–99461 (2019)
7. Vangelista, L.: Frequency shift chirp modulation: the LoRa modulation. IEEE Sig. Process. Lett. **24**(12), 1818–1821 (2017)
8. Mroue, H., Nasser, A., Parrein, B., et al.: Analytical and simulation study for LoRa modulation. In: 25th International Conference on Telecommunications (ICT), pp. 655–659. IEEE (2018)
9. Colavolpe, G., Foggi, T., Ricciulli, M., et al.: Reception of LoRa signals from LEO satellites. IEEE Trans. Aerosp. Electron. Syst. **55**(6), 3587–3602 (2019)
10. Ali, I., Al-Dhahir, N.: Doppler characterization for LEO satellites. IEEE Trans. Commun. **46**(3), 309–313 (1998)
11. Qian, Y., Ma, L., Liang, X.: The performance of chirp signal used in LEO satellite Internet of Things. IEEE Commun. Lett. **23**(8), 1319–1322 (2019)
12. Knight, M., Seeber, B.: Decoding LoRa: realizing a modern LPWAN with SDR. In: Proceedings of the GNU Radio Conference (2016)
13. Xhonneux, M., Bol, D., Louveaux, J.: A low-complexity synchronization scheme for LoRa end nodes. arXiv preprint arXiv:1912.11344) (2019
14. Ghanaatian, R., Afisiadis, O., Cotting, M., et al.: LoRa digital receiver analysis and implementation. In: IEEE International Conference on Acoustics, Speech and Signal Processing (ICASSP), pp. 1498–1502. IEEE (2019)
15. Bernier, C., Dehmas, F., Deparis, N.: Low complexity LoRa frame synchronization for ultra-low power software-defined radios. IEEE Trans. Commun. **68**(5), 3140–3152 (2020)
16. Reynders, B., Pollin, S.: Chirp spread spectrum as a modulation technique for long range communication. In: Symposium on Communications and Vehicular Technologies (SCVT), pp. 1–5. IEEE (2016)
17. Baofeng, Y., Yuehong, S.: The viterbi decoding algorithm of differential frequency hopping system with equivalent convolutional code structure. In: International Conference on Communications, Circuits and Systems, vol. 2, pp. 1117–1119. IEEE (2006)

Regular Space TT&C Mission Planning Based on Hierarchical Progress in Space Information Networks

Chenguang Wang[1], Di Zhou[1](✉) [ID], Min Sheng[1], Jiandong Li[1], and Yong Xiao[2]

[1] The State Key Laboratory of Integrated Service Networks, Xi'an 710071, Shaanxi, China
zhoudi@xidian.edu.cn
[2] The State Key Laboratory of Astronautic Dynamic,
Xi'an Satellite Control Center, Xi'an 710043, Shaanxi, China

Abstract. With the development of space missions and the diversity of satellites, space information networks have become more complex. Space TT&C (Tracking Telemetry and Command) missions are faced with shortage of resources and complex demand characteristics. In view of these challenges, we propose a hierarchical progressive TT&C mission planning algorithm. Firstly, we model the TT&C mission planning problem as a mixed integer linear programming problem with aiming at maximizing the network reward, i.e., the weighted number of completed TT&C missions. Then, we decompose this problem into multiple levels of optimization sub-problems according to the types of constraints and use time buckets and binary conflict trees to reduce the complexity of the algorithm. Finally, we use the binary conflict tree to perform local disturbances Simulation results show that, compared with the existing planning algorithms, the proposed algorithm not only guarantees the efficiency of the algorithm, but also improves the total reward of TT&C mission planning.

Keywords: Space information networks · Mission planning · Space TT&C

1 Introduction

Satellite network has unique advantages, and the space information network has always been regarded as a potential development direction. With the introduction of favorable national policies in the aerospace field and the application of many industrial-grade products, the total number of spacecrafts in orbit of china has exceeded 300 [1, 2]. It has an irreplaceable role in earth observation, emergency communications, air transportation and the expansion of national strategic interests [3–5]. As an indispensable part of space engineering, space TT&C (tracking telemetry and command) is responsible for

This work was supported in part by the Natural Science Foundation of China under Grant U19B2025, and Grant 62001347, in part by the China Postdoctoral Science Foundation under Grant 2019TQ0241 and Grant 2020M673344, and in part by the Fundamental Research Funds for the Central Universities under Grant XJS200117.

© Springer Nature Singapore Pte Ltd. 2021
Q. Yu (Ed.): SINC 2020, CCIS 1353, pp. 106–118, 2021.
https://doi.org/10.1007/978-981-16-1967-0_8

the adjustment of spacecraft's flight orbit and attitude, the control of on-board load and the monitoring of the working state of each module [6]. It is an important foundation to ensure the normal operation of spacecraft and complete the established task. In addition, the real-time monitoring of spacecraft can deal with emergencies quickly or early warning [7].

Because of the progress of satellite industry, the development of both military and civilian satellites is extremely rapid [8]. The types and functions of the on-orbit spacecraft are more abundant, and there are more requirements for ground equipment. Different ground stations are equipped with different functions, which increase the complexity of measurement and control tasks. In addition, due to the complexity of the international situation, the cost of establishing ground stations abroad is also high, so it is necessary to increase the utilization rate of ground measurement and control resources. In addition to the equipment factors, the task scale of spatial information network also shows an upward trend, and the complex mission demand and resource control also put forward higher requirements for the overall control ability of the measurement and control system [9]. In summary, with the expansion of spatial information networks and the complexity of tasks and equipment, the management of measurement and control systems must adopt timely and effective planning algorithms to integrate resources for planning. At present, the research on the routine measurement and control of spatial information network mainly includes:

Arbabi designed a two-stage non-conflict measurement and control request task number maximization method, which can reasonably describe the measurement and control task but has less constraint analysis on the equipment [10]. Jin proposed a nonlinear functional model, and simplified it into a 0–1 integer programming model with constraints under the assumption that the release stage affects the task planning, but this method ignores the constraints of the ground equipment on the task [11].

By using the method of graph theory, Zhang constructed the arcs and constraints into a graph, takes the shortest working time as the objective function, and transforms the resource scheduling problem into a subset optimization problem according to the working period [12]. Wang and Zhao used Petri network to analyze system elements such as requirements, equipment types, and actual dispatching rules to establish an operation model of a multi-satellite one-station system. But the scenario is simple and the subsequent demand arrangement is greatly affected by the previous arrangement [13].

By using machine learning and other algorithms, Liu built a constraint satisfaction model to maximize the total weight of the scheduling plan, and dynamically planned the task scheduling problem of the single-antenna ground station system [14]. Du divided the measurement and control resource scheduling system into task agents, network management agents and resource agents, and established a MAS model for space flight measurement and control task planning. However, the agent created by simple abstract rules cannot effectively describe different types of requirements and rules and there is still a big gap between the actual system [15].

Although there have been many studies on the planning of conventional measurement and control tasks, most of them have not added the complex constraints of the real system to the model, and lack consideration of the actual constraints of the measurement and control equipment. Therefore, in response to real constraints and complex requirements, we propose a multi-stage planning algorithm, which can use less time to obtain a higher task completion rate. The main contributions of the paper can be summarized as follows.

- We propose a conventional space TT&C mission planning based on the hierarchical progress of the spatial information network, which models the complex satellite and ground station constraints in the actual system, and decouples the related constraints according to the measurement. The constraints are divided into multiple stages for processing to ensure that the algorithm is consistent with the actual situation. This reduces the degree of constraint coupling and be conducive to the implementation of subsequent planning algorithms.
- Time bucket and arc conflict tree are established to ensure that iteration or arc replacement can be performed in a short time, so that the algorithm can be completed with lower time complexity.
- We propose a local perturbation algorithm based on the binary conflict tree ensures that the planning result will not get a locally optimal solution. Notably, the proposed algorithm can get a higher task completion rate through a limited number of iterations.

The rest of this article is as follows: The second part gives the network model and mathematical modeling of the problem. The third part transforms the model into a multi-level optimization problem, and solves it through the progressive thinking of levels. The fourth part is simulation, which is compared with related algorithms to verify the effectiveness of the algorithm.

2   System Model

2.1   Network Model

We consider a SIN scenario as shown in Fig. 1, which consists of several satellites, ground stations and a control center.

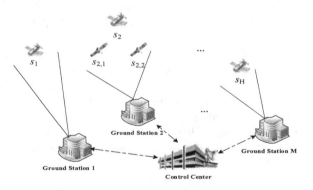

Fig. 1. Network scenario

Set $A = \{1, 2, \cdots, a, \cdots, M\}$ to represent the collection of ground stations. M represents the total number of ground stations, and each element represents a ground station. Each ground station has at least one available measurement and control equipment.

Satellite set is $S = \{S_1, S_2, \cdots, S_j, S_{j,1}, S_{j,2}, \cdots, S_H\}$ and H represents the total number of main satellites. Some main satellites have subordinate satellites, and satellites of the same constellation need to be observed together when observing the main satellite. For example, when observing main satellite S_j, also need to observe satellites $S_{j,1}$ and $S_{j,2}$ at the same time.

2.2 Measurement and Control Process

The mission of TT&C in this paper refers to the regular measurement and control task. Due to its periodic cycle, mission requirements will be delivered to the satellite management center at least one week in advance for unified planning.

A satellite measurement and control station is generally divided into task preparation stage, antenna preparation stage, measurement and control tracking stage and equipment release stage [16]. In this paper, the satellite forecast will give the visible arc segment of the satellite. In Fig. 2, it is shown that the antenna visible period of the satellite and the ground station (the ground station contains ground equipment) is established, where the antenna position time is fixed to 1 min, and the task preparation time is determined by the task master star, and the preparation time of the antenna and other equipment is determined by the equipment of the ground station.

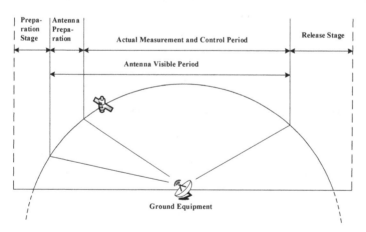

Fig. 2. Measurement and control task flow

2.3 Measurement and Control Requirements

The measurement and control requirements refer to the requirements and related constraints on equipment proposed by the task direction measurement and control center. It includes ground equipment location requirements, task location constraints, antenna elevation angle constraints, lifting rail requirements, lap arc requirements, and so on. These constraints can be integrated into related constraints based on the number of observation days. Measurement and control requirements are the overall input of conventional measurement and control task planning problems. The abstraction of constraints is difficult to quantify and cannot be directly used as the input of planning algorithms, and there are certain requirements for planning time.

2.4 Problem Model

Based on the network model proposed in Sect. 2.1 and the measurement and control requirements proposed in Sect. 2.3, this section will construct the problem model.

Measurement and control tasks have certain requirements for ground equipment and satellites. Each device of the ground station contains information such as the time of building and dismantling the chain, the function of the device, frequency band, signal system (service) support, and the time period of prohibition. The satellite arc forecast includes the corresponding satellite code and ground station code, including the satellite circle number and the number of transit circles and total transit circles obtained through calculations, the status of the arc up and down orbit and key point information.

In addition to equipment requirements, there are also some requirements for the measurement and control task itself. According to the different constraints, it can be divided into two types: this circle constraint and inter-circle constraint: this circle constraint includes the lifting orbit requirement, the tracking elevation requirement, the task position constraint, the task circle type requirement, the equipment tracking mode constraint, and the equipment support function required; Inter-lap constraints include the constraints between different measurement and control laps of the same satellite. The constraints here are generally based on time or laps.

The task planning problem is modeled as a mixed integer linear optimization problem in which the objective function is to maximize the return value of the completed demand. As the constellation task is a future trend, the conditions for the completion of the constellation task are more demanding, so the completion rate of the constellation task should be guaranteed in the measurement and control task. In other words, according to the characteristics of different single-satellite missions, the average return value is w_1, and the average return value of all constellation missions is w_2, so there should be $w_1 < w_2$. Then, the optimization objective function of the planning algorithm is:

$$\max w_i \sum_{i \in S} \sum_{j \in D} \sum_{k \in K} x_{i,j}^k \tag{1}$$

S represents the set of satellites; D represents the set of mission days for satellite i; K represents the set of mission requirements for satellite i on the jth day; $x_{i,j}^k$ represents whether the kth requirement of mission satellite i on the jth day can be completed, with a value of 0 or 1. The constraints that need to be met are as follows:

$$\begin{cases} T_{\min} \leq T_{x_{i,j}^k} - T_{x_{i,j}^{k-1}} \leq T_{\max} \ k \geq 1 \\ T_{\min} \leq T_{x_{i,j-1}^{K_l}} - T_{x_{i,j}^k} \leq T_{\max} \ k = 0 \end{cases} \tag{2}$$

$$ET_{dis} \geq ST_{x_{i,j}^k}, ET_{x_{i,j}^{k-1}} \leq ST_{dis} \tag{3}$$

$$F_{x_{i,j}^k} \subseteq F_D \tag{4}$$

$$x_{i,ii,j}^k = 1, ii \in L_i \tag{5}$$

Formula (2) gives that the task needs to meet the inter-circle constraints, and T_{min} T_{max} represents the task time or the number of tasks. K_l is the last measurement and control task of the satellite on the previous day. If the arc segment is not the first arc segment of the day, it needs to be a certain distance from the previous task arc segment of the day (circle or time). Formula (3) ensures that any valid task directly does not conflict with the device disabled time period. ET_{dis} represents the disabled time period end time set, ST_{dis} represents the disabled time period start time set, $ET_{x_{i,j}^k}$ represents the current arc end time set, $ST_{x_{i,j}^k}$ represents the current arc start time set. Formula (4) ensures that the satellite and equipment functions selected by the effective mission arc are adjusted and supports the corresponding vertical pitch angle and lifting orbit requirements, where $F_{x_{i,j}^k}$ represents the function orbit extension support list required by $x_{i,j}^k$, and F_D represents the support list of equipment. Equation (5) guarantees that if the mission is a constellation mission, it is necessary to ensure that all secondary stars can complete the observation mission in the same circle. The secondary satellite observation is divided into two types according to the requirements, whether to observe with the same equipment or to coordinate with multiple equipment Observe, among which L_i is the secondary satellite list of mission satellite i, and ii represents one of the secondary satellites. In the above optimization model, variable $x_{i,j}^k$ is 0–1 variable. Observing Eqs. (2)–(5), it can be found that the constraints are all time-varying constraints, and the non-time-varying constraints can be dealt with in the process of input preprocessing.

3   Hierarchical Progressive Programming Algorithm

Due to the complex constraints and strong relevance, in the actual planning process, this paper uses hierarchical progressive planning for step-by-step planning. Since the demand constraints of planning are roughly divided into two categories, one is the constraint that has nothing to do with the time circle, that is, the fixed constraint, and the other is the immediate change of the constraints that need to be located on the task day and the circle. Therefore, pre-process the measurement and control requirements and solve the former before the planning starts (Table 1).

The preprocessing algorithm mainly consists of two parts. The first part is the processing of fixed constraints. The update of the start and end time of the forecast arc will be added to the set of tasks to be selected and added to the time bucket. The time bucket records the arc set of each time interval. In this way, when calculating arc conflicts in the second part, the complexity of the algorithm can be reduced from O(n2) to O(m * n), where m is the arc of the designed time bucket, which is much less than the total arc (n) (Table 2).

Table 1. Preprocessing algorithm

1: Obtain the satellite forecasts, ground station information, and measurement and control mission requirements;

2: Circulate ground station information and measurement and control task requirements, and integrate the requirements into *demandMap* ;

3: Circular satellite forecast, generate *possibleTa skMap* ;

4: *for(possibleTa sk ∈ possibleTa skMap)*{

5: According to *demandMap* , the ground station information adds the construction and demolition chain of the arc time window and the task preparation time;

6: Determine whether the fixed constraint conditions are met, remove the arcs that do not meet the constraints, and update the start and end time of the arcs that meet the requirements and add them to the time bucket;

7: }

8: *for(task ∈ possibleTa skMap.keySet())*{

9: *for(possibleTa sk ∈ possibleTa skMap.get (task))*{

10: Compute conflict tree;

11: }

12: }

13: Get the final *possibleTa skMap* .

According to Algorithm 1, after the preprocessing, the fixed constraint has been processed, and a conflict tree is added to each arc to be selected. It stores the arcs of the remaining mission satellites that conflict with the current arc in a binary tree structure. For example, as shown in Fig. 3, if arc 1 is the current mission arc, and arc 2 is the arc of the remaining mission satellites, it will be placed on the right subtree of the root node on the conflict tree, otherwise it will be placed on the left subtree. If there is a containment relationship, it will be added to the root node's containment relationship conflict list. The conflict tree can be used to reduce the operation of deleting conflicting arcs from $O(n)$ to $O(1)$ in the subsequent planning.

Algorithm 2 is used for the initial planning after the preprocessing, because the fixed mission satellite measurement and control tasks on the same day include the requirements for changing orbital changes and requirements that do not require the orbital-changing requirements. Therefore, the task requirements for the orbit change are planned first. If the task is successfully completed, the current mission satellite will be added to the candidate arc in proportion, otherwise only one candidate will be entered into the queue. The priority queue here is implemented with a small top heap, and the one at the top of the heap always has the smallest conflict tree size. After processing the above tasks, the remaining tasks are enqueued to obtain the final initial planning result (Table 3).

Table 2. Initial programming algorithm

1: Input preprocessing algorithm output results;

2: $for(d \leq D)\{$

3: Calculate the total number of orbit change tasks in d and record it as $countCur$;

4: $for(task \in possibleTaskMap.keySet())\{$

5: If $task$ is an orbit change task on day d , add $C_{i,d} \div n_{i,d}$ feasible arcs of feasible task arcs to the priority queue pq according to the number of tasks $n_{i,d}$ before the orbit change and the number of arcs $C_{i,d}$ available before the orbit change;

6: $while(pq ! = \mathbf{Null}, countCur > 0)\{$

7: Obtain an arc from the top of pq for time-varying constraint judgment.

8: If the result is true, delete the arc set corresponding to the conflict tree and update the conflict tree;

9: Reorder pq and update the inter-circle demand and task completion status in the $demandMap$;

10: }

11: }

12: Clear pq and plan the remaining tasks (including the tasks after the orbit change and the tasks that have not been changed) according to the above-mentioned process again;

13: Obtain planning results $successTasks$.

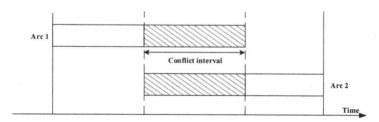

Fig. 3. Schematic diagram of arc conflict

After completing the initial planning of the task, it can be found that this algorithm is a time-based linear programming strategy, which results in that the arc selected each time is a local optimal solution, so the initial planning algorithm must be further processed. First traverse the failed task set, add optional arcs to the planning queue according to the proportion of uncompleted tasks, perform local disturbances, stop the disturbance if there is no better solution within the specified number of tasks, and output the optimal solution.

Table 3. Local disturbance algorithm

1： Enter the maximum number of iterations N, the initial planning result; *successTas ks* ;
2： $for(d \leq D)\{$
3： $for(de \in demandMap)\{$
4： If the number of unfinished tasks in de is 0, *continue*;
5： Re-plan and delete conflict arcs according to the conflict tree
6： Update *successTas ks* ;
7： }
8： }
9： Output the final *successTasks* .

4 Simulation

4.1 Simulation Parameters

We consider the following five scenarios: the total number of satellites includes all primary and secondary satellites, and provides arc forecasts for all ground stations. The total number of tasks is the number of main satellites to be planned. The main satellites here are divided into two categories, one is the main satellites with the constellation task of the secondary satellites, and the other is the task without the secondary satellites. Each task contains 9–13 days of measurement and control task requirements. The number of days here is not a natural day, as described above. Each day contains up to four measurement and control requirements, including variable lift and drop rail requirements and no requirements for lifting rail, etc. Therefore, the requirement is the sum of all the days of the main star. The total number of forecasts is the total number of arcs in the measurement and control mission time of the satellite, where the arc forecast contains all the main and secondary satellites (Table 4).

Table 4. Simulation parameters table

Scene	Total number of satellites	Total number of missions	Total number of demands	Total number of forecasts
A	198	170	5637	159993
B	199	171	5557	165978
C	199	169	5530	165261
D	204	174	5404	147188
E	205	175	5738	197751

4.2 Simulation Results

In order to verify the effectiveness of Mission Planning Algorithm Based on Hierarchical Progress (MPABHP) for spatial information network designed in this paper, the planning algorithm proposed in this paper is compared with the commonly used greedy algorithm (GA), random avoidance algorithm (RAA) and queuing algorithm for calculating conflict priority (QA). The realization of the greedy algorithm is the greedy algorithm based on the end time of the time window. The random avoidance algorithm is an iterative algorithm based on the greedy algorithm. The queuing algorithm for conflict priority is a planning algorithm that counts the number of conflicts between an arc and the rest of the arc as a priority, and performs a constraint judgment after sorting according to this priority.

IntelliJ IDEA is used to implement the algorithm, and the advantages of the designed measurement and control algorithm are analyzed by comparing the number of tasks completed by the two algorithms and the total income of the completed tasks in these five scenarios.

Comparison of Task Completion of Each Algorithm
The results of the algorithm in this paper, greed, random avoidance, and conflict priority fulfillment requirements and total requirements are shown in Fig. 4.

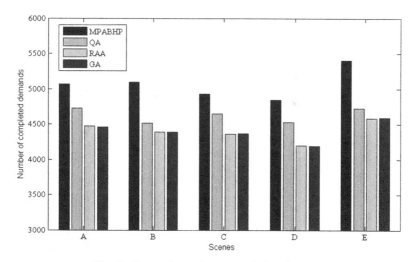

Fig. 4. Comparison of task completion demands

In the five scenarios, the algorithm in this paper has certain advantages in terms of the total number of completed requirements. Greedy based on the end time of the time window is the optimal solution for unconstrained time window planning, but because the measurement and control tasks share ground equipment and there are many resource couplings, the greedy algorithm does not bring a high demand completion rate. Random evasion algorithm is an iterative method of greedy algorithm, but because the measurement and control task has too many time-related constraints, it is not very effective to perform disturbance or iteration based on the initial solution of time greedy. The conflict degree priority algorithm does not have as many requirements as the algorithm proposed in this paper. This is because the algorithm in this paper considers the priority of processing the change of track requirements, so that more resources are used to give greater demand for subsequent impact. A binary conflict tree is introduced in the algorithm to ensure that the algorithm can find the conflicting arc faster, update the parameters, and perform neighborhood disturbances more targeted. In summary, the hierarchical progressive algorithm can increase the number of requirements to be completed for this type of measurement and control task with rich constraints.

As shown in Fig. 5, for the constellation task requirements among the requirements, the completion of the constellation task requires a lot of resources, which is not worthwhile for the greedy and random avoidance algorithms, so they complete few constellation tasks. In the conflict priority algorithm, because the constellation is considered in the priority setting, the number of completed constellation tasks has increased to a certain extent, and the conflict of constellation tasks is still more than that of single-satellite tasks, so the number of completed tasks is not significantly improved. As for the hierarchical progressive algorithm, due to the hierarchical planning, more resources may flow to the constellation demand, so the completion amount of the constellation demand is higher. In summary, the hierarchical progressive algorithm guarantees a higher degree of completion of the constellation requirements under the premise of ensuring a high degree of task completion.

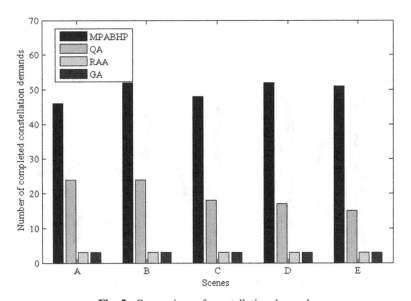

Fig. 5. Comparison of constellation demands

Comparison of the Total Income of Each Algorithm
In different task scenarios, the total reward of the tasks completed by the four algorithms is shown in Fig. 6, and the time consumption is shown in Fig. 7. The algorithm designed in this paper has a higher return value under the premise that the time complexity is much lower than the comparison algorithm. In some scenarios, the conflict priority algorithm pays too much attention to the total number of tasks, and will select some requirements with lower profit value but low conflict degree to complete first. In this

scenario, the advantages of the algorithm proposed in this paper are more obvious, and the high demand is maintained. Under the premise of completion degree, a high overall return value is also guaranteed. The gap between random avoidance and greed in terms of return value is still very small, which shows that simple iteration cannot improve the return value of greedy planning.

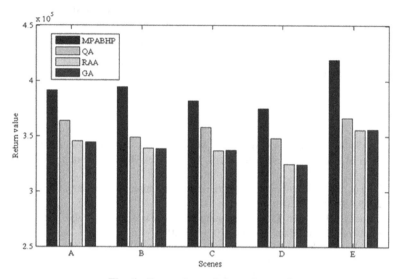

Fig. 6. Comparison of planned reward

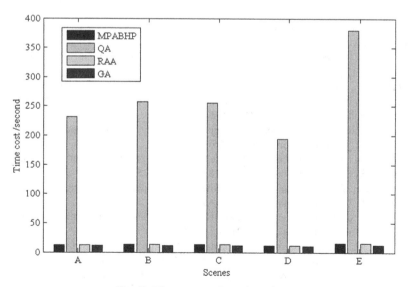

Fig. 7. Time-consuming comparison

5   Conclusion

Aiming at the increasingly complex measurement and control planning problem, this paper designs a hierarchical progressive planning algorithm. Firstly, the planning problem is modeled as an optimization problem that maximizes the value of reward. Secondly, using the hierarchical approach to split constraints, the modeled global problem is split into multiple levels of time window planning problems. The final planning set is obtained by solving the sub-problems of each layer solution, and then the global optimal solution is obtained through local disturbance, and the final scheduling scheme is generated. Finally, simulations verify that the algorithm in this paper has gains in the number of tasks completed and the total reward.

References

1. Wang, Y., Sheng, M., Zhuang, W., et al.: Multi-resource coordinate scheduling for earth observation in space information networks. IEEE J. Sel. Areas Commun. **36**(2), 268–279 (2018)
2. Song, C., et al.: New development of satellite communication in the 5G Era. Digital Commun. World **2019**(08), 36–37 (2019)
3. Liu, R., Sheng, M., et al.: Tasking planning based on task splitting and merging in relay satellite network. J. Commun. **38**(S1), 110–117 (2018)
4. Li, B., et al.: The development status of broadband satellite networks. Electron. World **2012**(13), 127 (2012)
5. Hao, C., et al.: Challenges for the development of satellite internet network. Digital Commun. World **2019**(07), 35–36+49 (2019)
6. Li, J.: coordinated task planning research of space-aeronautics earth-observing. Doctor, National University of Defense Technology (2013)
7. Chen, X.: Research on the acquisition technologies in spread spectrum TT&C system. Doctor, National University of Defense Technology (2010)
8. Yi, C., et al.: The recent development and prospect of satellite communication. J. Commun. **36**(06), 161–176 (2015)
9. Clement, B.J., Johnston, M.D.: The Deep Space Network scheduling problem. In: Twentieth National Conference on Artificial Intelligence & the Seventeenth Innovative Applications of Artificial Intelligence Conference. DBLP (2005)
10. Arbabi, M., Garate, J.A., Kocher, D.F.: Interactive real time scheduling and control summer. In: Computer Simulation Conference (1985)
11. Jin. G., et al.: Ground station resource scheduling optimization model and its heuristic algorithm. Syst. Eng. Electron. **2004**(12), 1839–1841+1875 (2004)
12. Zhang, N., et al.: A new model for satellite TT&C resource scheduling and its solution algorithm. J. Astronautics **30**(05), 2140–2145 (2009)
13. Wang, Y., et al.: Study on petri net model for multi-satellites-ground station system. J. Air Force Eng. Univ. (Nat. Sci. Edn.) **4**(02), 7–11 (2003)
14. Liu, Y., et al.: The method of mission planning of the ground station of satellite based on dynamic programming. Chin. Space Sci. Technol. **2005**(01), 47–50 (2005)
15. Du, H.: Design of space TT & C resource scheduling model based on multi agent technology. Syst. Simul. Technol. Appl. **16**, 78–81 (2015)
16. Li, Y.: Research on satellite-ground station data transmission scheduling models and algorithms. Doctor, National University of Defense Technology (2008)

Mobility Management in Low Altitude Heterogeneous Networks Using Reinforcement Learning Algorithm

Yunpeng Hou, Chao Wang, Huasen He$^{(\boxtimes)}$, and Jian Yang

University of Science and Technology of China, Hefei, China
hehuasen@ustc.edu.cn

Abstract. Unmanned aerial vehicle (UAV) base station has been proposed as a promising solution in emergency communication and supplementary communication for terrestrial networks due to its flexible layout and good mobility support. However, the dense deployment of UAV base station and ground base station brings great challenges in the configuration of neighbor cell list (NCL) during handover process. This paper presents a Cascading Bandits based Mobility Management (CBMM) algorithm for NCL configuration in the low altitude heterogeneous networks, where online learning is used to exploiting the historical handover information. In addition to the received signal strength, the cell load of each base station is also considered in the handover procedure. We aim at optimizing the configuration of NCL, so as to improve handover performance by increasing the probability of selecting the best target base station while at the same time reducing the selection delay. It is proved that the signaling overhead can be effectively reduced, since the proposed CBMM algorithm can significantly cut down the number of candidate base stations in NCL. Moreover, by ranking the candidate base stations according to their historical performance, the number of measured base stations in handover preparation phase can be effectively reduced to avoid extra delay. The simulation results of the proposed algorithm and other two existing solutions are presented to illustrate that the CBMM algorithm can achieve efficient handover management.

Keywords: Handover management · Heterogeneous networks · Mobility management · Online learning algorithms

1 Introduction

With the development of wireless communication technologies, the demand of high-quality network service increases dramatically. Under the current 5G heterogeneous network architecture, macro base stations (MBSs) provide basic coverage for ground users, while small base stations provide high speed and low latency network service. However, in crowded urban areas, due to some special events, such as concerts, large celebrations and so on, traditional cellular networks cannot provide good quality of service because of their limited communication capacity. The deployment of base stations in these regions has become saturated, and increasing the density of base station will

© Springer Nature Singapore Pte Ltd. 2021
Q. Yu (Ed.): SINC 2020, CCIS 1353, pp. 119–131, 2021.
https://doi.org/10.1007/978-981-16-1967-0_9

boost the complexity for network management while the network performance may be degraded due to the increase of interference. Moreover, consider the time and space uncertainty of special events with ultra-high traffic demand, it is not cost-efficient to deploy new ground base station. Therefore, deploying aerial base station (ABS) on UAVs to facilitate mobile communication is regarded as an effective scheme [1] to improve the quality of service of terrestrial networks.

Because of the flexibility and maneuverability, UAV base station can provide wider coverage compared to ground base station and alleviate the congestion of downlink traffic. A low-altitude network with low cost, high equipment reusability and good mobility compatibility can realize service diversion in these areas thanks to the merits of UAV including high flexibility and convenient deployment. Thus, the low altitude heterogeneous network has a promising future and can be widely used in different scenarios [2, 3]. Furthermore, an appropriate deployment of UAV base stations as a supplementary communication mode for the existing terrestrial cellular networks, can provide a better Line-of-Sight (LoS) propagation path for ground devices to improve the communication quality. The deployment of UAV base stations can achieve auxiliary service for cell edge users and effectively improve the edge throughput performance of cellular cells [4]. However, the high-density deployment of various types of base stations brings some challenges to network management, while handover management is a critical issue [5, 6].

Handover management enables network service for user devices not to be interrupted during movement, which ensures the continuity of service. In the mobile handover protocol, pilot channel quality of adjacent base stations is measured and reported to the serving base station by the mobile user equipment (UE). To improve handover performance, researchers from industry and academic have carried out many related works on handover management. For example, in [7], to reduce the cost of operation and maintenance in wireless networks and improve handover performance, the Automatic Neighbor Relations (ANR) scheme was proposed. Based on ANR, the target handover base station should be selected from a group of specific candidate base stations with high channel quality. To prevent handover failure, the serving base station configures a specific set of adjacent base stations, namely Neighbor Cell List (NCL), and pushes them to the UE. At this point, the length of NCL and the channel quality of base stations in NCL can significantly influence the performance of mobility management [8]. To optimize handover performance in high-density wireless networks, NCL configuration is considered as an important task. On the one hand, eliminating unnecessary base stations in NCL will reduce signaling overhead and handover delay, while the energy consumption of handover can also be decreased. On the other hand, NCL should contain enough candidate base stations, so that there exists one or more base stations that can be selected as target base station and handover failure can be avoided.

Mobility management in wireless networks has become a hot topic in recent years, especially in the process of NCL configuration. There has been some existing works in the literature. At the initialization phase of a network, all cells overlapped with the serving base station can be selected according to the network topology to realize manual NCL configuration. Static information including base station locations, antenna direction and receiving signal directions is adopted to forecast cell coverage and neighbor relationship.

For instance, different weighting factors are assigned to the static parameters of a base station and NCL is manually configure by using decreasing adjacent coefficients in [9]. Furthermore, the configuration of NCL can be optimized by collecting relevant parameters during the operation of the network. At the initial stage, a base station needs to sense the spectrum and detect all the base stations near it, then the key problem in the measurement stage is whether the NCL should include the detected neighbor base station. Considering the statistics of previously accessed cells and the distance estimation between base stations, [10] presented a theory of blocking paths to assist user to detect adjacent base stations.

Although the existing NCL configuration methods can play a certain role in traditional wireless networks, it will encounter many challenges when facing the special characteristics of the low altitude heterogeneous network which integrates UAV base stations. Firstly, the existing handover management schemes does not take into account the impact of cell loads in UE handover procedure. Compared with ground base station, the capacity and coverage of UAV base station are limited, and it is sensitive to the number of accessed user devices [4, 11]. Secondly, because of the difference of geographical environment between UAV base station and ground base station, the channel characteristics of communication between them and terminals are quite different. Since there are few obstacles for UAV, the propagation path between UAV base station and ground user is mainly LoS [12]. Finally, the existing NCL configuration scheme only considers instantaneous system performance without reference to prior knowledge such as long-term performance estimation of base stations and historical handover results [13]. Therefore, in this work, we propose a reinforcement learning based mobility management method which takes both the base station load and historical handover information into consideration.

2 System Model

2.1 Network Model

In this paper, a low altitude heterogeneous network model which integrates terrestrial network and UAV base station is considered, in which UAV is used as aerial base station to provide supplementary coverage and necessary data transmission for ground UE. When natural disasters or crowd-intensive public events occur, the ground base stations cannot provide sufficient network service. By deploying UAV base stations, emergency communication networks or supplementary communication networks are established to ensure the normal communication service for users. The network architecture is shown in Fig. 1. In this work, base stations with different power and coverage are considered, including macro base stations $BS_{mac} = \{b_1^m, b_2^m, \ldots, b_l^m\}$, micro base stations $BS_{femto} = \{b_1^p, b_2^p, \ldots, b_m^p\}$ and UAV base stations $BS_{uav} = \{b_1^u, b_2^u, \ldots, b_n^u\}$. We assume all base stations are randomly ordered and denoted as $K = \{BS_{mac}, BS_{femto}, BS_{uav}\}$, and each type of base stations use the same transmit power while different types of base stations have different transmit power, that is $P_{mac} > P_{femto} > P_{uav}$ [4]. To reflect the actual communication scenario, we use a weighted Voronoi graph to fit the coverage of the ground base stations and set the average density of base station at 128–512/km^2 [14].

In addition, the user distribution in the network obeys an independent Poisson point process with density λ_P, and the number of users in a particular cell obeys the Poisson distribution with mean value $\overline{N} = \lambda_P S$, where S is the coverage area of the cell.

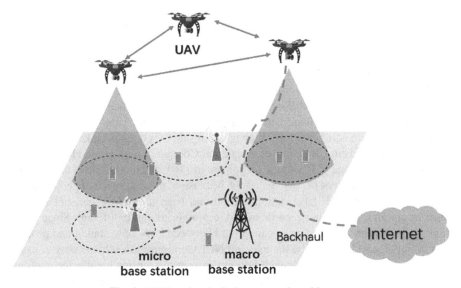

Fig. 1. UAV-assisted wireless network architecture

In the communication scenario of UAV-assisted cellular network, we aim at solving the handover problem faced by UE in the process of movement. As the UE moves within the coverage of its serving base station, the channel quality is continuously monitored and the measurement report is sent to the serving base station. Furthermore, the connection between base stations through Xn interfaces enables direct communication between base stations to assist user devices to handover between different base stations. 15 A handover process will be triggered when the measurement report provided by the UE satisfies the A1–A6 or B1–B4 event [15]. For each base station, we adopt the received signal strength to reflect the channel quality, and the received signal strength of each candidate base station should not less than -70 dBm [12]. In addition, the cell load is defined as the ratio of the number of served UE to the maximum number of serviceable UE at the same time, which should not exceed 3/4 [11].

In the proposed mobility management scheme based on reinforcement learning, the handover process is disposed in a time slot, and the learning is completed by iterations in multiple time slots. The handover process generally includes three stages: 1) handover preparation, 2) handover execution, and 3) handover completion. The handover preparation stage which configures a NCL for UE is our main target. When a handover is completed, the reward within time slot t is calculated and the performance of the calculated NCL is evaluated according to the handover index, such as the signaling overhead and handover delay of scanning candidate base stations. Then the operation is repeated in the next time slot $t + 1$. The iteration process is presented in Fig. 2.

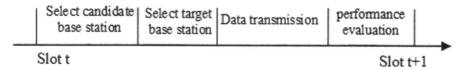

| Select candidate base station | Select target base station | Data transmission | performance evaluation |

Slot t   Slot t+1

Fig. 2. Learning iteration for mobility management based on reinforcement learning

2.2 Problem Formulation

The optimal handover strategy needs to minimize the NCL length and optimize the base stations sequence in the NCL. For more accurate evaluations of base station, the channel quality and cell load should be considered simultaneously in NCL optimization.

Since the transmission environment of ground base stations and UAV base stations is different, different path loss models should be adopted for different types of base stations. Consider the influence of urban environment on the channel between ground base station and mobile UE, we follow the settings of HetNet scenarios in 3GPP standard and use 3GPP path loss model in small cellular networks [16] which can be expressed as:

$$PL(d)[dB] = 15.3 + 37.6 \times log_{10}(d) + PL_{ow}, \quad d > d_0 \tag{1}$$

Based on the Air-to-Ground (A2G) channel model presented in [17], the LoS and non-light-of-sight (NLoS) links between a UE and its serving UAV base station are considered, and the average path loss of the A2G link can be written as:

$$PL(h, d) = PL_{Los} \times Pr_{LoS} + PL_{nLoS} \times Pr_{nLoS} \tag{2}$$

Here, Pr_{Los} and PL_{Los} represent the LoS link probability and path loss respectively, which are defined as:

$$Pr_{LoS}(r, h) = \frac{1}{1 + a \exp\left\{-b\left(arctan\left(\frac{h}{d}\right) - a\right)\right\}} \tag{3}$$

$$PL_{Los} = 20 \log\left(\frac{4\pi f_c d}{c}\right) + \eta_{LoS} \tag{4}$$

For NLoS link, the link probability and the path loss are given as:

$$Pr_{nLoS}(r, h) = 1 - Pr_{LoS}(r, h) \tag{5}$$

$$PL_{LoS} = 20 \log\left(\frac{4\pi f_c d}{c}\right) + \eta_{nLoS} \tag{6}$$

where f_c denotes the carrier frequency and d represents the horizontal distance between UEs and its serving UAV base station, h is the height of the UAV and η represents the mean value of the multi-path loss.

For the handover UE, $P_{i,t}$ represents the received signal strength of the i-th base station, and $P_{i,t}$ is relevant to the small-scale fading as well as the distance-dependent path loss at time slot t. Rayleigh fading is used to describe the channel fading in this work.

For each time slot, $P_{i,t}$ is an unknown random distributed sample value. The expected value of $P_{i,t}$ can be expressed as

$$\overline{P}_i = P_{BS} - PL_{LoS} = \lim_{T \to \infty} \frac{1}{T} \sum_{t=1}^{T} P_{i,t} \tag{7}$$

where P_{BS} represents the transmission power of different base stations.

$\eta_{i,t}$ is the cell load of base station i at time slot t, which represents the proportion of the number of served UE to the maximum cell capacity. Due to the random distribution of mobility UE, uniformly distributed Poisson Point Process (PPP) is used to describe the UE location. For the i-th base station, the number of connected UE obeys a Poisson distribution, and the mean value of the distribution is positively correlated with the number of devices in the base station coverage area. The upper limit of UE served by UAV base station is only 1/6 of that of ground macro base station [8]. The average cell load of the i-th base station is defined as

$$\overline{\eta}_i = \lim_{T \to \infty} \frac{1}{T} \sum_{t=1}^{T} \eta_{i,t} \tag{8}$$

Combining with reinforcement learning algorithm, for time slot t, $reward_t$ represents the reward value of learning agent, which measures the performance of the NCL L_t calculated at time slot t. Assuming that $\widetilde{L}_t$ is the set of base stations measured before the target base station is selected, we have $\widetilde{L}_t \subset L_t$. The handover target $L(|\widetilde{L}_t|)$ represents the $|\widetilde{L}_t|$-th element in the list. The cost of measuring base stations in L_t is proportional to the size of $\widetilde{L}_t$ and it is defined as $\beta|\widetilde{L}_t|$, where β is a constant. So, the $reward_t$ is defined as:

$$reward_t = \frac{P_{|\widetilde{L}_t|,t}}{\eta_{|\widetilde{L}_t|,t}} - \beta|\widetilde{L}_t| \tag{9}$$

3   Algorithm Design

3.1   Algorithm Description

To overcome the limitations of existing mobility management schemes, a more efficient NCL configuration algorithm is designed under the framework of low altitude heterogeneous network. Since user devices are able to learn the network environment and get accurate performance information about the base station, we consider reinforcement learning algorithms with decision-making capabilities, which interact with the environment in a trial-and-error manner. The optimal strategy is obtained by maximizing the cumulative reward. A Multi Armed Bandits (MAB) mathematical model is used in this algorithm to formulate the base station selection problem in mobility management. The mobility management problem faces the tradeoff between using the current target base station with the highest sampling value and trying to select the base station with less sampling times. Rewards calculated at different time slot are independent of each other. Moreover, the reward is determined by the unknown distribution of the received signal strength and the cell load. The basic idea of Bandits algorithms for estimating unknown distributions comes from the Upper Confidence Bound (UCB) Strategy, which is associated with each item in the behavior set by calculating the UCB index. For a specific project, UCB index depends on the reward and cost sequences generated throughout the learning process.

To solve the problem of NCL configuration during handover, we propose a Cascading Bandits based Mobility Management (CBMM) algorithm based on the UCB strategy, which includes initialization phase and iteration phase. In the initialization phase, UE moves to the cell edge of the serving base station and triggers handover. It's serving base station will configure a NCL for the UE. The UE scans and measures all base stations in NCL until a base station which meets all handover requirements is found. For the i-th base station, UE measures its received signal strength P_i and checks the base station load η_i. After finding the target base station and completing the handover, these measurements and other indexes (e.g. the scanned base station number) will be reported and stored in the Mobility Management Entity (MME).

In the iteration phase, MME appraises the base stations in a NCL according to the historical measurement information and calculates the corresponding UCB index. The algorithm will rank the base stations whose UCB indexes are not less than the threshold according to the descending order of the UCB indexes. In the t-th iteration, the calculation result of the current time slot can be obtained and recorded as L_t. UE scans each base station in L_t until the handover target is found. MME updates the performance evaluation of the base stations according to the reported sample values of the received signal strength and the cell load of each measured base station, then the corresponding reward value is calculated according to Eq. (9). In the $(t+1)$-th iteration, the learning process is repeated when a UE moves to the same position and triggers a handover.. The optimal NCL for a specific location can be obtained after T iterations. After that, a UE moves to that location can use the results directly, thus avoiding repeat or redundant calculations.

The pseudo code of the proposed CBMM algorithm is presented in Algorithm 1. The total time slots that UE selects the k-th base station before time slot t is expressed as $N_{k,t}$. The sample values of P_k and η_k at the t time are expressed as $\hat{P}_{k,t}$ and $\hat{\eta}_{k,t}$ respectively. The algorithm uses $u_{k,t} = \sqrt{\frac{\alpha \log t}{N_{k,t}}}$ to calculate the UCB index of the k-th base station at time slot t, where α is a constant not exceed 1.5.

Algorithm 1 Cascading Bandits based Mobility Management Algorithm

Input: iteration time T; constant coefficient α ; UCB threshold τ

Output: NCL

Init: Randomize the initial parameters of base stations in the network, including receiving signal strength and cell load;

Iterate: while t<T:

    For k = 1, 2, ..., |K|:

        If $\frac{\hat{P}_{k,t}+u_{k,t}}{\hat{\eta}_{k,t}+u_{k,t}} > \tau$ then:

        $k \rightarrow L_t$

    Rank the base stations in L_t in the descending order which satisfied $\frac{\hat{P}_{k,t}+u_{k,t}}{\hat{\eta}_{k,t}+u_{k,t}} > \tau$

        For $i = 1: |L_t|$:

            Measure the received signal strength and cell load of the i-th base station

            If the i-th base station satisfies handover requirements:

                break

        Update parameters, $N_{i,t}$, $\hat{P}_{i,t}$, $\hat{\eta}_{i,t}$;

        Calculate $reward(t)$ and $cost(t)$

        $t = t + 1$

3.2 Algorithm Analysis

For Bandits theory, the cumulative regret value in finite time is generally used as the key index to measure the convergence of the algorithm. The regret value is defined as the difference between the reward of the iterative result in the learning process and the reward value corresponding to the theoretical optimal solution. So we calculate regret(t) = $reward^* - reward(t)$, where $reward^* = \frac{\overline{P}_{i^*}}{\overline{\eta}_{i^*}} - \beta$ represents the reward corresponding to the theoretical optimal solution L^* calculated under the condition that the global information is known. The optimal result L^* covers the base stations that meet the condition $\frac{\overline{P}_i}{\overline{\eta}_i} > \tau$, and all the base stations are ranked in descending order based on $\frac{\overline{P}_i}{\overline{\eta}_i}$, which means the optimal base stations i^* (may not be the handover target base station) are at the top of the optimal list L^*.

In this section, the finite time regret value is used as a function of time to analyze the performance of Algorithm 1, and the upper bound of cumulative regret value is provided. We set $\Delta_i^2 = (\overline{P}_i - \overline{\eta}_i)^2$, and the regret value for a given time T is:

$$R(T) \leq \sum_{i \in K \backslash L^*} \frac{\{(16K\alpha)log(T)\}}{\Delta_i^2} + O(1) \tag{10}$$

It can be noticed that the additional cost corresponding to the number of measurements has no impact on the order of the upper limit of the cumulative regret value, and the sublinear relationship between the cumulative regret value and the number of T iterations ensures the convergence of the algorithm. As the learning results of the algorithm gradually converge to the best NCL, the extra overhead due to the unnecessary base station scanning gradually decreases and approaches to 0.

4 Experimental Results

4.1 Experimental Settings

In this work, we use MATLAB to build the simulation environment. The target area is set up as one $1000\,m \times 1000\,m$ urban area, where 4 macro base stations, 64 micro base stations and 36 UAV base stations are deployed. The macro base station is randomly distributed with a minimum distance $r_{min} = 300\,m$ between each other, the distance from the micro base station to its nearest macro base station is not less than 50 m, and the UAV base station is set at a low altitude of 100 m. The coverage radius of UAV is 100 m, while UAVs are randomly deployed in the target area. The transmit power and other parameter settings of different types of base stations are shown in Table 1:

Table 1. Simulation parameters of a low altitude heterogeneous network

Simulation parameters	Reference value
Macro station transmit power P_{mac}	35 dBm
Micro station transmit power P_{femto}	23 dBm
UAV base station transmit power P_{uav}	20 dBm
Thermal noise density	-174 dB/Hz
Carrier frequency	2.1 GHz
Penetration loss	10 dB

Moreover, the UE in the network environment is distributed on the basis of a Poisson Point Process while the intensity λ is set to $1/250$. It is assumed that each UE of the considered region moves in random direction to leave or enter the coverage of a specific base station. We assume that a specific UE located at the cell edge of the connected base station triggers handover. We study the handover process of the UE and takes its results as a representative of the overall system performance. The Rayleigh fading and statistical path loss model are utilized to calculate the received signal strength for the target UE and it is assumed that the cell load of each base station is linearly related to the number of UEs in its coverage region. The upper limit of load capacity of UAV base station is set to 200 UEs, while the load capacity of micro-base and macro-base stations is set to 1.5 times and 6 times of that of UAV base station [4, 8]. In each iteration of the learning process, the locations of UE are generated randomly. During the network initialization phase, we assume that each UE is served by the base station which provides the highest received signal strength.

4.2 Comparison Algorithms and Performance Indicators

The CBMM algorithm is compared with the other two existing schemes. The first comparison algorithm is a NCL configuration scheme which adopts dynamic threshold and indicated as "DT-based solution" [18]. In DT-based solution, candidate base stations are selected according to the RSS measured by the UE and the probability of handover to each base station, so that the base station which frequently becomes the handover target will be more likely to be selected. To implement the DT-based solution, the handover probability of a specific base station is calculated as the statistics of 10^3 historical results. The other comparison algorithm is the "RSS-based solution" [7]. In RSS-based solution, NCL is determined based on the instantaneous performance of the candidate base station, and base stations whose sample value of RSS is higher than the handover threshold will be added to the NCL [7].

To compare the handover performance of different solutions, this paper selects two indicators: the NCL length, and the number of base stations before the handover target is selected. NCL length is proportional to the signaling overhead during handover, and the scan base stations number significantly affects the handover delay in the preparation phase.

4.3 Experimental Results

In our simulations, the proposed CBMM algorithm is implemented and compared with two existing solutions. Experiment 1 and experiment 2 study the handover performance of the three solutions in terms of NCL length and the number of scanned base stations. Experiment 3 evaluates the efficiency of the CBMM algorithm by comparing the performance of the NCL obtained after different iteration times.

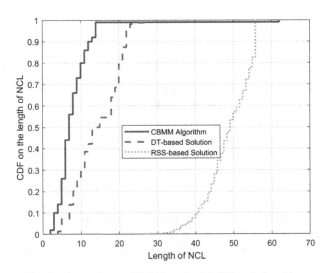

Fig. 3. Comparison of NCL lengths for different algorithms

First, Fig. 3 shows the comparison of cumulative distribution functions of NCL length obtained under different algorithms Here, we use "CBMM algorithm", "DT-based solution" and "RSS-based solution" respectively to represent the proposed algorithm and two comparison algorithms. The results in Fig. 3 is obtained from 500 independent exams. In Fig. 3, it can be observed that the NCL length of the CBMM algorithm is less than that of the two comparison algorithms. Consider the average number of candidate base stations in each solution, it can be noticed that the averaged NCL length obtained by CBMM algorithm is reduced by nearly 30% and 80% of that in the DT-based and RSS-based solutions. This is because the density deployment of different types of base station resulting in a large number of candidate base stations that meet the RSS requirement. However, some overloaded base stations should not be added to NCL. The proposed CBMM algorithm can significantly reduce the NCL length by avoiding adding high load base stations to NCL.

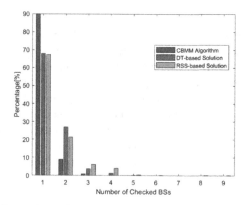

Fig. 4. Comparison of the number of measured base stations in NCL

The statistical results of the number of measured base stations in handover process for different algorithms are illustrated in Fig. 4. Since the measurement of unnecessary base stations in NCL will cause additional delay in the handover process, experimental results show that the proposed CBMM algorithm is advantageous to the other two contrast algorithms in reducing the number of measured base stations. Especially, the probability that the first base station in a NCL is selected as the target base station is greater than 80%. However, the probabilities for DT-based and RSS-based solutions are only 68% and 66% respectively. The average number of measured base stations corresponding to CBMM algorithm is 1.11, while the average numbers for DT-based and RSS-based algorithms are 1.39 and 1.47 respectively. This is because the learning-based algorithm considers both the channel conditions and the cell load of base stations, and the evaluation of candidate base stations is more accurate. For traditional RSS-based solution, instantaneous RSS is used to evaluate candidate base stations, while the long-term performance of base stations is neglected. Considering the number of scanned base stations, the CBMM algorithm provides better performance than the existing solutions, and it can significantly reduce the preparation delay by accurately estimating the performance of the base stations and adjusting the order of candidate base stations in the NCL.

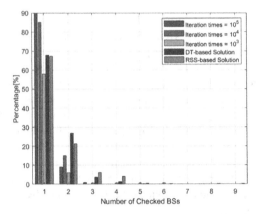

Fig. 5. Performance analysis for different iteration times

Finally, experiment 3 evaluates the efficiency of the proposed algorithm by setting different learning iterations. It can be observed from Fig. 5 that when the number of learning iterations is set to 10^3, 10^4 and 10^5, the mean number of measured base station decreases. We show that the proposed algorithm outperforms the comparison algorithms when the learning algorithm iterates 10^4 times. Notably, the performance of the CBMM algorithm begins to degrade when the number of iterations is less than 10^3. The intuitive interpretation of this result is that the distribution of RSS and the learning accuracy of the cell load depend on the number of iterations. Hence, long-term performance estimates for candidate base stations are important in NCL configuration.

5 Conclusions

This work focused on the handover management in low altitude heterogeneous wireless networks where UAV base stations were deployed to assist terrestrial networks. We focus on the NCL configuration and optimization in handover preparation phase. Different from existing solutions where instantaneous performance of base stations are adopted in NCL configuration, we use online learning framework to exploiting long-term performance estimation in the handover decision process. We have proposed a CBMM algorithm to optimize NCL when handover is triggered. Simulation results shown that the length of NCL can be significantly reduced by adopting the CBMM algorithm while the number of measured base stations is also cut down during the handover preparation phase. We show that our proposal can decrease the excessive signaling overheads as well as the delay in handover preparation phase, which was proved to be an efficient mobility management scheme.

References

1. Lu, J., Wan, S., Chen, X., Fan, P.: Energy-efficient 3D UAV-BS placement versus mobile users' density and circuit power. In: 2017 IEEE Globecom Workshops (GC Wkshps), Singapore, pp. 1–6 (2017)

2. Cicek, C.T., Gultekin, H., Tavli, B., Yanikomeroglu, H.: UAV Base station location optimization for next generation wireless networks: overview and future research directions. In: 2019 1st International Conference on Unmanned Vehicle Systems-Oman (UVS), Muscat, Oman, pp. 1–6 (201)

3. Lai, C., Chen, C., Wang, L.: On-demand density-aware UAV base station 3D placement for arbitrarily distributed users with guaranteed data rates. IEEE Wirel. Commun. Lett. **8**(3), 913–916 (2019)

4. Alzenad, M., El-Keyi, A., Lagum, F., Yanikomeroglu, H.: 3-D placement of an unmanned aerial vehicle base station (UAV-BS) for energy-efficient maximal coverage. IEEE Wirel. Commun. Lett. **6**(4), 434–437 (2017)

5. Yin, S., Zhao, S., Zhao, Y., Yu, F.R.: Intelligent trajectory design in UAV-aided communications with reinforcement learning. IEEE Trans. Veh. Technol. **68**(8), 8227–8231 (2019)

6. Xiao, Z., Dong, H., Bai, L., Wu, D.O., Xia, X.: Unmanned aerial vehicle base station (UAV-BS) deployment with millimeter-wave beamforming. IEEE Internet Things J. **7**(2), 1336–1349 (2020)

7. 3GPP, Evolved Universal Terrestrial Radio Access (E-UTRA) User Equipment (UE) Procedures in idle Mode Version 15.3.0 Release 15, May 2019

8. Sharma, V., Bennis, M., Kumar, R.: UAV-assisted heterogeneous networks for capacity enhancement. IEEE Commun. Lett. **20**(6), 1207–1210 (2016)

9. Lv, Z., et al.: Neighbor cell list optimization of LTE based on MR. In: 2018 International Conference on Signal and Information Processing, Networking And Computers, Singapore, pp. 290–295 (2018)

10. Chowdhury, M.Z., Bui, M.T., Jang, Y.M.: Neighbor cell list optimization for femtocell-to-femtocell Handover in dense femtocellular networks. In: 2011 Third International Conference on Ubiquitous and Future Networks (ICUFN), Dalian, pp. 241–245 (2011)

11. Watanabe, Y., Matsunaga, Y., Kobayashi, K., Sugahara, H., Hamabe, K.: Dynamic neighbor cell list management for handover optimization in LTE. In: 2011 IEEE 73rd Vehicular Technology Conference (VTC Spring), Budapest, Hungary, pp. 1–5 (2011)

12. Yang, H., Hu, B., Wang, L.: A deep learning based handover mechanism for UAV networks. In: 2017 20th International Symposium on Wireless Personal Multimedia Communications (WPMC), Bali, pp. 380-384 (2017)

13. Shen, C., Tekin, C., van der Schaar, M.: A non-stochastic learning approach to energy efficient mobility management. IEEE J. Sel. Areas Commun. **34**(12), 3854–3868 (2016)

14. Wang, C., Yang, J., He, H., Zhou, R., Chen, S., Jiang, X.: Neighbor cell list optimization in handover management using cascading bandits algorithm. IEEE Access **8**, 134137–134150 (2020)

15. 3GPP TS 23.502. Procedures for the 5G System; Stage 2 (2019)

16. Telecommunication management; Configuration Management (CM); Notification Integration Reference Point (IRP); Requirements (2015)

17. Al-Hourani, A., Kandeepan, S., Lardner, S.: Optimal LAP altitude for maximum coverage. IEEE Wirel. Commun. Lett. **3**(6), 569–572 (2014)

18. Becvar, Z., Mach, P., Vondra, M.: Self-optimizing neighbor cell list with dynamic threshold for handover purposes in networks with small cells. Wirel. Commun. Mobile Comput. **15**, 1729–1743 (2015)

A Design Idea of Multicast Application in Satellite Communication Network

Jianglai Xu[✉], Jun Zheng, Lihong Lv, Zheshuai Zhou, and Yongshun Zhang

Beijing Space Information Relay Transmission Technology Research Center, Beijing 100094, China
mybird1234@sina.com

Abstract. IP multicast has the advantages of high network utilization, low bandwidth overhead and strong scalability, which plays a great role in a large number of terrestrial Internet applications. However, in satellite communication network, there are more applications for unicast environment, but less for multicast environment. The main reason is that the limited link bandwidth and high channel error rate restrict the application of multicast technology to a certain extent. Under the condition of satellite network bandwidth and multicast channel increasing, the performance of satellite communication channel is greatly improved. This paper analyzes the mode and protocol of network data transmission, and designs a method of distributing data files by using multicast.

Keywords: Satellite communication · Multicast · Retransmission · Response

1 Introduction

In the traditional terrestrial Internet, most of the applications are point-to-point transmission. Since the 1990s, in order to realize the application of resource sharing and multimedia conference, the point to multipoint IP multicast technology began to receive people's attention. Because the terrestrial Internet has the advantages of high network utilization, low bandwidth overhead, strong scalability and low bit error rate, IP multicast plays an important role in a large number of new network applications.

In the satellite communication environment, if multicast technology can be applied in the satellite communication network, it will further improve the resource utilization of satellite network and greatly expand the application field of satellite communication. But for a long time, the bandwidth of satellite communication link is limited, and the channel error rate is high, which is the bottleneck of restricting the application of multicast technology. With the development of satellite communication network, the performance of satellite communication equipment is improving, the link bandwidth is increasing, and the channel error rate is greatly reduced, which creates conditions and application possibilities for multicast applications in satellite communication network.

At present, the research on multicast technology has made great progress at home and abroad, but these researches mainly focus on the terrestrial Internet, the research on broadband satellite network mainly focuses on the unicast environment, and the research on multicast is less. Therefore, the application of multicast based on satellite network is a problem worthy of further study.

© Springer Nature Singapore Pte Ltd. 2021
Q. Yu (Ed.): SINC 2020, CCIS 1353, pp. 132–145, 2021.
https://doi.org/10.1007/978-981-16-1967-0_10

2   Analysis of Network Data Transmission

In satellite communication network, the usual way of data transmission is to establish a one-to-one TCP link between the sender and multiple receivers. The sender must copy multiple identical data files and transmit them to multiple receivers at the same time, which will lead to heavy burden, long delay, network congestion and low distribution efficiency.

As shown in Fig. 1, among the three receiving groups, group 1, group 2 and group 3, group 1 only expects data a, group 2 only expects data B, and group 3 only expects data C. group 1 has (3 receivers), group 2 has (3 receivers), and group 3 has (2 receivers). The file sender needs to establish one-to-one corresponding TCP connection with 3 + 3 + 2 receivers at the same time, and needs to copy 3 + 3 + 2 copies of ABC data to the server at the same time In the communication network, if the data capacity is large, it will cause long system delay, network congestion and low distribution efficiency. As shown in Fig. 2, if UDP multicast is used to distribute data files, and 3 + 3 + 2 data receivers join the same multicast address, the transmitter of satellite spot beam only needs to copy one copy of data and send it to the corresponding multicast address, and all data receivers in the reorganization can receive it, thus improving the efficiency of data file distribution.

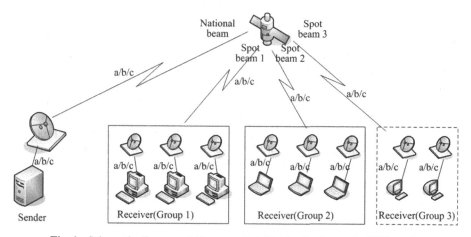

Fig. 1. Schematic diagram of telemetry data file distribution under TCP mode.

2.1   Network Data Transmission Mode

There are three ways of IP network data transmission: unicast, multicast and broadcast.

Unicast is the realization of point-to-point network connection between the sender and each receiver. Each receiver and the sender need an independent data channel. If a sender transmits the same data to multiple receivers at the same time, multiple copies of the same packet must be copied accordingly. If a large number of receivers want to get

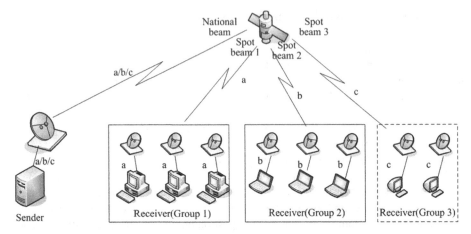

Fig. 2. Schematic diagram of telemetry data file distribution in multicast mode.

the same packets, it will lead to heavier link burden, long delay and network congestion. In order to ensure a certain quality of service, we need to increase the hardware and bandwidth, which is easy to do in the traditional terrestrial Internet.

Broadcast transmission is to broadcast data packets in the IP subnet, and all receivers in the subnet will receive these data packets. Broadcast means that whether these receivers need to receive the packet or not, the sender sends the same packet to each receiver in the subnet through the network. Therefore, the use of broadcast is very small, only effective in the local subnet, through the router and network equipment to control the broadcast transmission.

Multicast transmission means that when some receivers in the network need specific information, the multicast source (i.e. the sender of multicast information) only needs to send information once, and the multicast router uses the multicast routing protocol to establish a tree routing for the multicast data packets. The transmitted information is copied and distributed at the fork as far as possible, which greatly saves the network bandwidth and improves the efficiency of data transmission It solves the problem of low efficiency of unicast and broadcast.

2.2   Network Data Transport Layer Protocol

TCP (transmission control protocol) and UDP (User Datagram Protocol) are the two most important protocols in the transport layer. The main difference between UDP and TCP is that UDP does not necessarily provide reliable data transmission and can not guarantee the data to arrive at the destination accurately.

The purpose of TCP is to provide reliable data transmission and maintain a virtual connection between devices or services communicating with each other. TCP is responsible for data recovery when packets are received out of order, lost or destroyed during delivery. It completes this recovery by providing a sequence number for each packet it sends. The lower network layer treats each packet as an independent unit, so packets can be sent along completely different paths, even if they are all part of the same message.

This kind of routing is very similar to the way that the network layer deals with the segmentation and reassembly of packets, but only at a higher level.

To ensure correct data reception, TCP requires an ACK to be sent back when the target computer successfully receives the data. If the corresponding ack is not received within a certain time limit, the packet will be retransmitted. If the network is congested, this retransmission will result in duplicate packets. However, the receiving computer can use the sequence number of the packet to determine whether it is a duplicate packet and discard it if necessary.

If we compare the structure of UDP packet and TCP packet, it is obvious that UDP packet does not have the complex reliability and control mechanism of TCP packet. Like TCP protocol, the number of source ports and destination ports of UDP also support multiple applications on one host. A 16 bit UDP packet contains a byte long header and the length of the data. The check code field enables the overall check. In the case of UDP, even if the protocol knows that there is a broken packet, it will not be retransmitted.

The differences between TCP and UDP are: TCP is connection oriented; UDP is connectionless. TCP protocol requires the sender to shake hands three times to confirm whether the receiver receives the packet; UDP protocol does not. TCP protocol is slow, UDP protocol is fast.

When the link bandwidth is large and the channel bit error rate is low, TCP protocol is a natural choice, because the overhead caused by frequent handshake and reply can be offset by high-performance link bandwidth and channel bit error rate. UDP protocol is the best choice when the transmission performance is emphasized rather than the integrity of transmission. This is because there is no overhead caused by frequent handshake and response, and the efficiency of data transmission is greatly improved.

2.3   Application Layer File Protocol

In the application layer, for the data file distribution, if FTP is used for distribution, because FTP is based on TCP in the transport layer, it can achieve reliable transmission, but it is a single point to single point transmission mode. For multiple users to distribute and transmit synchronously, only multi line transmission can be adopted, which is bound to compete for system bandwidth, memory exchange, CPU and other resources, resulting in low efficiency.

If TFTP is used for distribution, because TFTP is based on UDP in the transport layer, it can complete the distribution and transmission to multiple users with only one task. However, the unreliability of UDP makes TFTP more suitable for small file transmission, because even if a small packet is lost or error occurs, the whole data transmission task may need to be executed again.

3   Key Technologies of Multicast

At present, the method of using UDP protocol to realize data multi-point distribution and transmission in ground communication network is very mature and has a wide application foundation. However, due to the asymmetry of satellite network and the inherent characteristics of single hop and double hop, such as long delay, high bit error rate, packet loss is a problem that we must face if we copy the UDP protocol of ground communication network to achieve multi-point data distribution. Therefore, under the condition of satellite communication network, it is necessary to improve UDP transmission protocol and add packet loss detection and retransmission mechanism.

3.1   Gateway Protocol Conversion

In the process of transmission, the transmission of information in satellite link is realized by Reliable UDP protocol. Reliable UDP protocol is implemented in the application layer, and the physical layer, link layer, network layer and transport layer are not affected. Therefore, the protocol kernel of the operating system does not need to be modified. It only needs to improve the ground network standard UDP protocol in the application layer to form a Reliable UDP protocol. Through the segmented protocol conversion module of the gateway, response control, data splitting and recovery, error detection and data recovery are added The retransmission function ensures the correctness of data transmission while ensuring the efficient transmission of network data. The gateway protocol conversion is shown in Fig. 3.

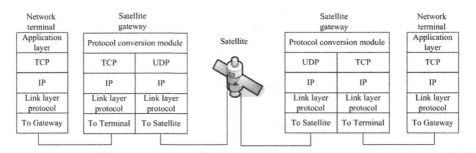

Fig. 3.   Gateway protocol conversion.

3.2   Response Control

Response control is implemented in the application layer and bound with UDP socket, including data flow response control and link detection information response control. Through response control, both ends of UDP communication become an organic loop. Application program can easily detect the on-off state of the link, know the receiving and sending situation of data in time, and execute data transmission according to response report and response timeout Retransmission control. Due to the asymmetric configuration of the transceiver link, and the return link rate is far lower than that of the high-speed transmission link data, in practical application, the full response control of the data stream is turned off, while only the full response process of the link detection is maintained, and the error response mechanism is adopted for the response of the data stream.

3.3 Data Splitting and Recovery

The ground communication network has good bit error rate characteristics. Data is transmitted between the computer terminal and the gateway in the way of connection oriented flow. In theory, the data packet can be infinite. Under the condition of satellite communication network, considering the characteristics of high bit error rate, long delay and retransmission, the size of UDP protocol data packet must be limited in a certain range, so as to ensure the full utilization of data When the channel quality is bad, the communication efficiency will not decrease obviously. Therefore, the data to be sent from the sender should be split, that is, the larger packets should be split into small packets suitable for satellite channel transmission. At the receiving end, the received small packets are recombined to recover the correct synthetic data.

3.4 Error Detection and Data Retransmission

The receiver receives the buffered data from the sender in turn from the multicast group address. When all the data are received, the receiver starts to check the data and records the results. The receiver checks and approves whether the total number of received packets is consistent according to the total number of packets in the frame information, and determines whether there is packet loss by combining with the packet number check. If there is packet loss, the receiver establishes its own packet loss queue. Error detection is judged by the check code in the received packet. When the check result is consistent, the frame transmission is completed. When the check result is inconsistent, the frame data is discarded and put into the retransmission queue.

Under the condition of high bit error rate satellite channel, the probability of data packet damage and loss in the transmission process is relatively large. In order to ensure the correctness of the final data, the receiver sends a retransmission request to the sender through the return channel immediately after receiving the wrong data or judging the frame loss. After receiving the return request, the sender terminates and records the current transmission processing, which needs retransmission After the transmitted data is sent, the previous transmission processing is performed. This reduces the number of retransmissions and the average end-to-end delay.

4 Design Idea of Multicast Application

Referring to the engineering application experience of the ground Internet, combined with the actual conditions of the satellite communication network, the improved UDP protocol multicast mode is adopted in the transmission layer, and the advantages and disadvantages of FTP and TFTP protocols are integrated in the application layer, so as to design a new method of distributing data files that can not only distribute at multiple points, but also ensure reliable transmission.

As shown in Fig. 4, the design idea of this new method is: to establish a multicast connection; the sender will packet the telemetry data file to be distributed, and then frame it according to the given packet length and frame format, and send it to the multicast address; the receiver will check the packet data and send a retransmission request to the sender on demand; the sender will check the receiver's ack information and retransmit it on demand; the file is divided into two parts After sending, close the multicast connection to standby mode. The workflow is shown in Fig. 5. The following describes each stage of work.

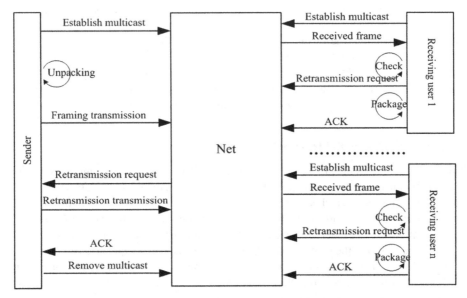

Fig. 4. Schematic diagram of data distribution method design.

4.1 Establishment and Removal of Multicast

The sender and the receiver of the file set up a multicast group according to the multicast address assignment, and each multicast group corresponds to a unique multicast address. The multicast group can be classified according to the file distribution range (that is, different combinations of the receiver). The sender and the receiver of the file are added to a multicast group by socket word processing, and the range of multicast address is 224.0.0-239.255.255.255. After the multicast group is established, all members of the group can carry out point-to-point data communication.

4.2 File Unpacking and Sending

Before sending the file, the sender will subcontract the telemetry data file to be distributed, divide it according to the given packet length and read it into the buffer. Each packet is framed according to the frame format shown in Table 1 and sent to the address in the multicast group.

The element fields of the file unpacking and framing format are as follows:

File name identification: the file name under the binary representation of the data file, with a value of 4 bytes.

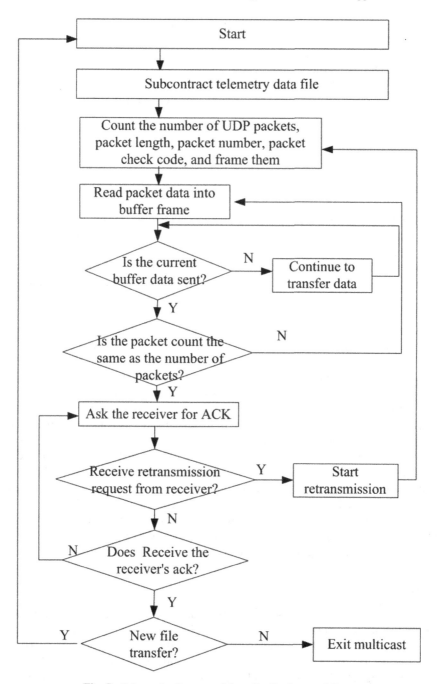

Fig. 5. Schematic diagram of data distribution workflow.

Table 1. Frame format of data file unpacking and framing.

File name identification	Source IP address	Multicast identification number	Split package length	Actual package length	Total contract number	Package number	Packet verification	Packet data
4 Bytes	4 Bytes	4 Bytes	4 Bytes	4 Bytes	4 Bytes	4 Bytes	2 Bytes	Variable, up to 1500 bytes

Source IP address: fill in the IP address of the sender to distinguish the source of the packet.

Multicast identification number: similar to the port, it identifies the unique multicast task with a value of 4 bytes. It includes syn, ACK, data and fin. There is no RECP data area for syn, ACK and fin.

Split package length: the maximum value of data file splitting and unpacking, which is selected between 0–1516 bytes, accounting for 4 bytes. According to the recommendation of windows system, the sum of packet length and other fields should not exceed 1516 bytes.

Actual package length: the value of the size of a single package corresponding to the unpacked data file, accounting for 4 bytes. The actual packet length may be less than the length of the split packet, for example, the data of the last packet after the split.

Total package number: the total package number of the data file disassembled according to the unpacking rules, with a value of 4 bytes. It can be obtained by modulus of data file size (bytes)/packet length (bytes).

Package serial number: the sequence code of disassembled data package, which is convenient for package data verification and data file assembly at the receiver, with a value of 4 bytes. The number of packets is counted from 1. After that, the number of packets will be increased by 1. The receiver uses this field to judge whether the packet is lost or out of order. The sequence number of the response packet is equal to the sequence number of the received packet, indicating that it is a response to this packet.

Packet check: check code, value takes 2 bytes, for example: MD5, CRC, etc.

Package data: the complete data in the data package disassembled by the data file according to the unpacking rules. The value is variable in bytes, up to 1500 bytes.

4.3   Verify Checksum

No matter how the size, format and quantity of data files change, MD5 can generate a unique "digital fingerprint" for any data file, which is usually called check code. If in the process of transmission, due to the loss of frames and packets, any changes to the data file will occur, the MD5 value, that is, the corresponding "digital fingerprint" - check code will change. In the sender, the check code is added to the information frame for transmission, while in the receiver, the check code is recovered. The received packet data is checked with windows md5 check once to ensure that the packet data obtained by the receiver and the packet data sent by the sender are the same data file. This is to use the check code to ensure the correctness of the received data.

Multiple receivers receive the buffered data from the sender in turn from the multicast group address. When all the data are received, the receiver starts to check the data and records the results. The receiver checks and approves whether the total number of received packets is consistent according to the total number of packets in the frame information, and determines whether there is packet loss by combining with the packet number check. If there is packet loss, the receiver establishes its own packet loss queue and sends the retransmission request of the sequential packet to the sender. After the sender receives it, the receiver retransmits the packet until there is no retransmission request. If the total package number and packet sequence number received by the receiver are normal, the check code in the current round of buffer will be determined by the check code in the received packet, Then the verification data is taken out and compared with the verification results calculated by ourselves. When the data is consistent, the verification is completed and the current round of transmission is completed. When the data is inconsistent, the error is reported to the sender and the whole buffer is requested to be retransmitted. In addition to receiving data, writing to memory, checking calculation and other operations, the receiver can also make an active response at an appropriate time. By adding logical judgment on the rationality of the actual packet length, multicast identification number, and packet sequence number, the receiver discards redundant or disordered packets, and notifies the sender of the resend request in time when necessary.

4.4 Packet Retransmission

After receiving the retransmission request from the receiver, the sender recovers the data packet corresponding to the sequence packet from the retransmission request and retransmits the packet to the multicast group address. Each receiver in the multicast group checks according to their own needs. When the sequence number data packet has been received and passed the verification, the sender actively discards the packet; when the sequence number data packet has not been received or the sequence number has not been received If the packet fails to pass the verification, the sequence number packet will be retained. This loop will continue until the sender receives all the acks from the receiver. In order to prevent the program from dead loop caused by long check in the transmission process, an exit check mechanism is established for the number of retransmissions of the sequence number packet.

4.5 File Package Composition

After receiving the packets from the sender, each receiver, after the logical judgment of the actual packet length, multicast identification number, packet serial number, total package number, and the check and verification of packet data, recombines the packets according to the reverse order of the sender's unpacking rules, and forms a new data file locally. After the completion of this process, it sends ACK message to the sender, removes the multicast connection, enters the standby mode, and waits for the next data file distribution and reception. Although the ACK information is very small and the possibility of transmission loss is also very small, in order to ensure the integrity and

reliability of the file distribution process, the sender should establish an ACK information inquiry mechanism to deal with the small probability event of the receiver's ack
information loss.

5 Desktop Simulation Test

In order to verify the effect of multicast application design under the condition of satellite
communication network, we consider using TL-1 relay satellite system to build a test
environment. There are two kinds of test environment design considerations, one is to
use real satellite to forward space link, the other is to use wired frequency conversion to
simulate satellite to forward space link. The former needs to occupy the actual satellite
system resources, and the test coordination cost is relatively low High. In the latter, the
desktop test simulation environment can effectively use the maintenance gap of the relay
satellite system, and the test simulation cost is low, which can achieve the purpose of
verifying the multicast application design under the condition of satellite communication
network.

The desktop test simulation environment is shown in Fig. 6. The ground terminal
station uses the forward modulator of the medium and low speed terminal to access the
gateway equipment, the gateway equipment to access the Ethernet switch, the Ethernet
switch to access a computer terminal to simulate a forward user s, the simulation test
station uses the backward demodulator of the medium and low speed terminal to access
the gateway equipment, the gateway equipment to access the Ethernet switch, and the
Ethernet switch to access five computer terminals To simulate one to five backward
users (R1, R2, R3, R4, R5). The test scenario is that the forward user s of the ground
terminal station transmits the file data to the medium and low speed forward modulator
in turn through the data processing module after the TCP (UDP multicast idea) packet.
The signal of the up conversion link of the ground terminal station to the antenna out-
let coupling link is sent to the SSA forward backward analog repeater (equivalent to
the function of the on-board repeater), and the SSA forward backward analog repeater
generates the downlink after frequency conversion Then it is coupled to the down conversion link of the simulation test station, and then it reaches the middle and low speed
terminal backward demodulator of the simulation test station, and then it reaches one to
five simulated backward users (R1, R2, R3, R4, R5) through the gateway and Ethernet
switch. Five scenarios can be described as follows:

Scenario 1 s > R1 (1 user to 1 user)
Scenario 2 s > R1, R2 (1 user to 1 user)
Scenario 3 s > R1, R2, R3 (1 user to 3 users)
Scenario 4 S > R1, R2, R3, R4 (1 user to 4 users)
Scenario 5 s > R1, R2, R3, R4, R5 (1 user to 5 users)

The data files to be distributed are 996 kb (about 1 MHz), 100032 kb (about
100 MHz), 499121 kb (about 500 MHz) and 999811 kb (about 1 GHz). Record the
time required for each file distribution and transfer in each scenario.

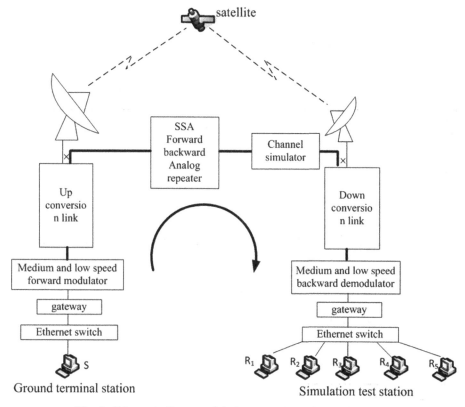

Fig. 6. Schematic diagram of desktop test simulation environment.

Each computer terminal runs UDP multicast distribution file simulation module (as shown in Fig. 7) to complete the sending and receiving of data files and the time-consuming record, in which the data packet and packet combination are completed by the simulation module program.

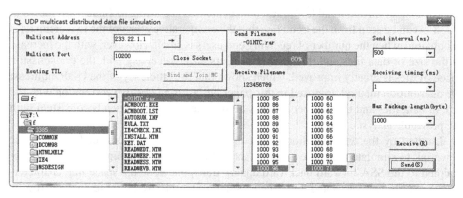

Fig. 7. Desktop test simulation environment.

The simulation results of five scenarios are shown in Table 2. In the table, T represents the transmission time, size represents the file size, M represents the distribution method, and UDP-m represents the UDP multicast.

Table 2. Time comparison test results of two ways of distributing data files.

Scene classi-fication	Scenario 1		Scenario 2		Scenario 3		Scenario 4		Scenario 5	
t \ m \ s	TCP	UDP-m	TCP	UDP-m	TCP	UDP-m	TCP	UDP-m	TCP	UDP-m
1 MHz	1s	2s	1.5s	1.5s	2s	2s	4.5s	5s	7s	3s
100 MHz	14s	16s	15s	15s	18s	20s	40s	30s	52s	35s
500 MHz	101s	85s	110s	90s	115s	95s	145s	125s	195s	150s
1 GHz	205s	192s	210s	195s	230s	185s	270s	190s	350s	210s

In order to intuitively analyze the effect of file transmission speed in five scenarios, the larger the difference Δt is, the better the transmission efficiency of the latter is. The time difference between the two methods is shown in Table 3.

Table 3. Time difference results of data file distribution in two ways.

Δt \ m \ s	Scenario 1	Scenario 2	Scenario 3	Scenario 4	Scenario 5
1 MHz	-1	0	0	-0.5	4
100 MHz	-2	0	-2	10	17
500 MHz	16	20	15	20	45
1 GHz	13	25	45	80	140

The histogram and trend line of test data generated according to Table 3 are shown in Fig. 8. It can be seen from the trend line that the comparative effect of the two methods is not obvious when the number of receiving end users is small. However, with the increase of the size of data files and the number of users at the receiving end, the transmission speed of files is greatly improved, which shows that the efficiency of the UDP multicast file distribution method proposed in this paper has a strong positive correlation with the size of data files and the number of users at the receiving end, which is consistent with the previous analysis.

Limited by the test conditions, only the channel simulator is added to the space downlink to simulate the space link characteristics. The space uplink signal is directly coupled to the SSA forward backward analog transponder, so the space link characteristics are only half simulated, but this does not affect the overall trend analysis results of the test data.

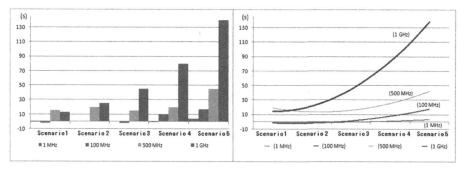

Fig. 8. Generated histogram and trend chart of test data.

6 Concluding Remarks

Under the condition of satellite communication network, compared with ground communication network, there are obvious differences in channel environment and topology. Desktop simulation is an ideal simulation. Due to the difference between satellite communication network and ground communication network, the data splitting size of reliable multicast must be considered in the tradeoff between data throughput, delay performance and cost. However, with the development of satellite communication network, the performance of satellite communication equipment is improving, the link bandwidth is increasing, and the channel error rate is greatly reduced, which will also create application conditions for multicast applications in satellite communication network.

References

1. Wu, Q.: Set up the eternal lifeline - emergency communication system based on TD-SCDMA. Mob. Commun. 4 (2009)
2. Sun, B., Geng, S., Yin, S., et al.: Application mode of relay satellite S frequency band multiple access system. Spacecr. Control Measur. **35**(1), 81–84 (2016)
3. Pan, L., Du, P.: Study on real-time performance of data relay transmission. J. Aircr. Control **30**(5), 63–66 (2011)
4. Feng, H., Chen, Y., Wu, X.: Satellite ground station resources allocation SVM regression model. Control J. **30**(2), 33–35 (2011)
5. Xin, Y., Zhou, X.: Media access control protocol for broadband GEO satellite network. Commun. Technol. **48**(12), 106–109 (2010)
6. Zhang, X., Wang, R., Dong, X., et al.: New Technology and System Application of Emergency Communication. China Machine Press, Beijing (2009)
7. Cao, G.: Suggestions on the development of heaven and earth integrated emergency communication. In: 2010 National Emergency Communication Seminar, pp. 51–59 (2010)
8. Min, S.: Current situation and prospect of emergency communication development in China. In: 2010 National Emergency Communication Seminar, pp. 60–64 (2010)
9. Xue, C., Sun, F.: Design and implementation of data monitoring software based on PDXP protocol. Flight Control Measur. **30**(6), 39–42 (2011)
10. Zhou, X., Wang, Z.J., She, Y.: Analysis of delay performance of CFDAMA-PR MAC protocol for satellite rotation reservation. Comput. Appl. Res. (2006)

Research on Power Control Algorithm in the Multi-target TT&C System

Lihong Lv[1](✉), Yiwen Chen[1], Dandan Fan[1], Jianglai Xu[1], and Zheshuai Zhou[2]

[1] Unit 32039 of PLA, Beijing, China
[2] Unit 32011 of PLA, Beijing, China

Abstract. In the multi-target TT&C system of relay satellite, due to the different distances between users and satellites, the strong signals of near users will restrain the weak signals of long-distance users and reduce the system capacity. This paper studies the power control algorithm in the reverse link of multi-target TT&C system. Specifically, the signal model of multi-target TT&C system is established, and two power adjustment algorithms, namely fixed step adjustment algorithm and adaptive step adjustment algorithm, are analyzed. The adaptive step adjustment algorithm is improved, and the upper limit of step adjustment is introduced. The simulation results show that the algorithm is effective in multi-target TT&C system.

Keywords: Multi-target TT&C system · Power control · Adaptive step length · Relay satellite

1 Introduction

The multi-target TT&C system of relay satellite used orthogonal address codes distinguish multiple users, which has many advantages: (1) Low signal power spectral density, when covering the entire time domain and frequency domain, its power spectral density is much lower than FDMA and TDMA methods, and it has Good signal concealment. (2) Large system capacity, the CDMA system is an interference-limited system, and reducing interference can bring about an increase in system capacity. (3) It can be integrated with pseudo code ranging. Pseudo code ranging calculates the distance by measuring the phase of the pseudo random code. And CDMA also uses PN codes to distinguish different users, so integrating pseudo code ranging and CDMA can realize data transmission and pseudo-code ranging at the same time. However, CDMA also has some disadvantages: (1) Because the distance between the user and the multi-target TT&C system is different, the strong signal of close users will inhibit the weak signal of far users. (2) Due to the asynchronous problem of different user signals in the actual system, the address codes are no longer completely orthogonal, which leads to interference between different users. Both of these problems can be summarized as multiple access interference problems. It is necessary to study anti-multiple access interference technologies to improve system capacity. The existing anti-multiple access interference technologies can be summarized into the following categories [3]:

© Springer Nature Singapore Pte Ltd. 2021
Q. Yu (Ed.): SINC 2020, CCIS 1353, pp. 146–157, 2021.
https://doi.org/10.1007/978-981-16-1967-0_11

(1) Address code design, the system's ability to resist multiple access interference is directly affected by the cross-correlation performance between different codes. (2) Channel coding technology, in the message sequence, reduce bit error rate by adding redundant information according to the coding algorithm. This technology is widely used in single-target TT&C systems, but it does not use the relevant characteristics of interference, but treats the interference as noise, which has a limited increase in capacity. (3) Multi-user detection technology, which no longer treats interference as noise, but treats it as information with a certain structure [4]. Specifically, the correlation between address codes is used to suppress or even eliminate the influence of multiple access interference. (4) Power control technology means that each user adjusts the transmission power according to certain criteria so that the power reaching the central station is equal, which can eliminate the power difference between different users caused by the near-far effect, but it cannot compensate for the capacity loss caused by non-orthogonal address codes.

The above four technologies, address code design and channel coding technology have been very mature and widely used in various practical systems. Power control technology and multi-user detection technology, however, are new technologies brought about by multi-user access, and have always been a hot research field of CDMA, but current research is mainly concentrated on applications in terrestrial mobile communication systems, lacking relay satellite multi-target measurement and control applications Research under the background, this is the main content of this article.

2 Signal Model of the Multi-target TT&C System

The multi-target TT&C system uses spread spectrum measurement and control technology. Common spread spectrum methods include Direct Sequence Spread Spectrum (DSSS), Frequency Hopping (FH), and Time Hopping (TH),Broadband Chirp Modulation (Chirp) and hybrid spread spectrum, etc. [8] Among them, DSSS is the most convenient to implement. The aircraft spread spectrum measurement and control system uses this method to achieve spread spectrum. The system block diagram is shown in Fig. 1. The spread spectrum system introduces spread spectrum and despread modules. At the sender, the original information is modulated and then spread. Spreading is to modulate the information to be transmitted and the PN sequence to achieve spectrum spread, while the despreading at the receiving end uses the same PN sequence demodulation information, spread spectrum and despread use modulo-2 sum operation.

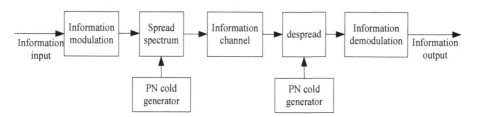

Fig. 1. Block diagram of spread spectrum measurement and control system

In multi-target measurement and control, the channel is no longer exclusive, but shared by multiple users. CDMA assigns different PN codes as address codes for each user. Common PN codes include m sequence, Gold sequence and Walsh sequence. The m-sequence correlation performance is outstanding, but the number of available sequences is too small to meet the needs of multiple users. In the case of synchronous transmission, the performance of Walsh sequence is very well, but in the case of asynchronous transmission, the performance deteriorates seriously, The multi-target TT&C system uses the Gold sequence, which has a large number of sequences, can fully meet the needs of multiple users, and the related performance is pretty good.

The signal of The multi-target TT&C system is shown in Fig. 2, $r_k(t)$ indicates the signal sent by each user, $n(t)$ represents additive white Gaussian noise, $r(t)$ represents the signal received by the user. The signal of the kth user is shown in formula (1):

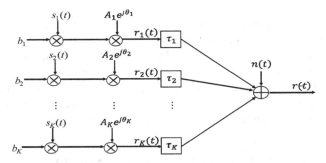

Fig. 2. Signal diagram of multi-target measurement and control system

$$r_k(t) = A_k s_k(t) b_k(t) e^{j\theta_k} \tag{1}$$

Among them, A_k represents the signal amplitude received by the kth user, $s_k(t)$和$b_k(t)$respectively represents the spreading code waveform and data waveform of the kth user, θ_k represents the phase difference (including the instantaneous value of the frequency deviation) of the kth user, τ_k represents the delay of the kth user. According to whether the system is synchronized, the value of θ_k and τ_k are different.

The multi-target TT&C system is an asynchronous CDMA system, that is, the carrier frequency and time are asynchronous. A typical asynchronous CDMA system is a return link for multi-target measurement and control. Each user independently generates a return signal, so there is a carrier phase deviation θ_k, and the signal arrives at the relay satellite separately, so the delay τ_k is also different. The expression of the signal $r(t)$ received by each aircraft is as follows:

$$r(t) = \sum_{k=1}^{K} r_k(t - \tau_k) + n(t) \tag{2}$$

Among, $r_k(t) = A_k s_k(t) b_k(t) e^{j\theta_k}$.

The channel characteristics used in wireless communication will change with time and frequency. According to the different effects of the channel on the signal in time and frequency, there are two types of fading, large-scale fading and small-scale fading. The former is usually not related to frequency, including path loss and shadow fading, while the latter is related to frequency, including delay spread and Doppler spread [2]. For large-scale fading, according to the electromagnetic wave propagation formula, the signal strength will gradually decrease as the distance increases, and this intensity loss can be characterized by path loss, the path loss formula [1] is as follows:

$$L = 32.45 + 20\lg(f) + 20\lg(d) \tag{3}$$

Among them, the unit of path loss L is dB, the unit of frequency f is MHz, and the unit of distance d is km.

During the flight of the aircraft, the speed constantly changes, which causes the received frequency to also change, which is called the Doppler effect [1]. The resulting frequency offset is called Doppler frequency offset, and the calculation formula is as follows:

$$f_D = \frac{v}{\lambda}\cos\varphi = \frac{vf}{c}\cos\varphi \tag{4}$$

Among them, v is the relative flight speed, f is the carrier frequency, c is the speed of light, the angle between the flight direction and the beam arrival direction.

The Doppler frequency offset of different paths is different, and Doppler spread can be used to characterize the time scale of channel changes, which can be described by Doppler power spectrum, typically the probability density function introduced by Jakes, as shown below:

$$P_{f_D}(f_D) = \begin{cases} \dfrac{1}{\pi f_{D\max}\sqrt{1-(f_D/f_{D\max})^2}}, & |f_D| < f_{D\max} \\ 0 & , \quad \text{others} \end{cases} \tag{5}$$

Among them, $f_{D\max}$ means Maximum Doppler deviation.

The Jakes model is aimed at each multipath evenly distributed in 360°, but for aircrafts, the multipath's arrival angle is concentrated in a fixed narrow interval. This interval is called the beam width and is usually selected as 3.5°, and this Doppler probability spectrum is a segment of the intercepted Jakes spectrum:

$$P_{f_D}(f_D) = \begin{cases} \dfrac{1}{(\varphi_{\alpha_H}-\varphi_{\alpha_L})f_{D\max}\sqrt{1-(f_D/f_{D\max})^2}}, & f_{D\max}\cos\varphi_{\alpha_H} < f_D < f_{D\max}\cos\varphi_{\alpha_L} \\ 0 & , \quad \text{others} \end{cases} \tag{6}$$

Among them, φ_{α_H} 和 φ_{α_L} respectively represent the maximum angle and minimum angle of beam arrival.

3 Power Control Algorithm of the Multi-target TT&C System

Firstly, power control enables each target to be locked by the tracking module of the receiver, and then adjusts the transmit power to achieve optimal power consumption. The objective function is as follows:

$$\Delta_{SINR} = |SINR_{obs} - SINR_{th}| < \Delta \tag{7}$$

Setting a $SINR_{th}$ according to the demodulation threshold of the receiver, and then adjusting the transmit power, aims to make the difference between the $SINR_{obs}$ and $SINR_{th}$ of each observed aircraft less than Δ, The specific process is shown in Fig. 3. The transmission power control in the whole process can be divided into two stages: increasing the transmission power before locking and adjusting the transmission power after locking. The former is relatively simple, it is enough to increase the transmit power continuously with a fixed step length, and the latter needs to design an adjustment strategy to achieve rapid adjustment. This article mainly studies the adjustment algorithm of transmit power.

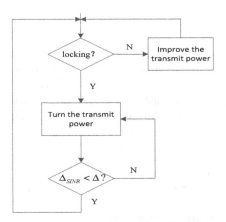

Fig. 3. Power control flow chart

3.1 The Fixed Step Adjustment Algorithm

The simplest adjustment algorithm is based on a fixed step to adjust the transmit power, which is called a fixed step adjustment algorithm. Assuming that the adjustment step is δ, the fixed step adjustment algorithm uses the following strategy to adjust the transmit power of each aircraft:

$$P(n+1) = P(n) - \delta \, \text{sgn}(SINR_{obs} - SINR_{th}) \tag{8}$$

Among them, $P(n)$ is the current transmit power, $P(n+1)$ is the adjusted transmit power, and the units of $P(n)$, $P(n+1)$ and δ are all dB. The flow of the algorithm is shown in the Fig. 4 below:

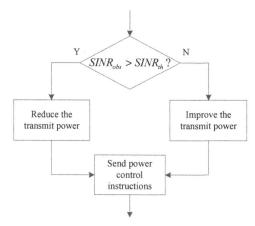

Fig. 4. Flow chart of fixed step adjustment algorithm

Adjustment algorithm are as follows:

(1) The relay satellite observes the signal-to-interference and noise ratio of each aircraft $SINR_{obs}$ and compares it with the target signal-to-interference and noise ratio $SINR_{th}$;

(2) If $SINR_{obs} > SINR_{th}$, means that the transmitting power of the aircraft is too large, and it is required to reduce the transmitting power by δ. Otherwise, it means that the transmission power is too small, and it is required to increase the transmission power by δ;

(3) The relay satellite generates a power control command and sends it to each aircraft. Each aircraft adjusts the transmission power according to the power control command, and waits for the system to respond to the power change before starting the next iteration.

As shown above, the fixed step adjustment algorithm is very simple and easy to implement, and because the step length is fixed, the power control instruction content only needs 1 bit, but when the difference between $SINR_{obs}$ and $SINR_{th}$ is large, the convergence time of the algorithm is too long.

3.2 Adaptive Step Size Adjustment Algorithm

In view of the slow convergence speed of the fixed step adjustment algorithm, there have been many variable step adjustment algorithms. One type of adjustment step size is adaptively changed, which is called an adaptive step size adjustment algorithm. The adjustment strategy of its transmitting power is shown in the following formula:

$$P(n+1) = P(n) - \delta(n)\text{sgn}\left(SINR_{obs}^{n} - SINR_{th}\right) \tag{9}$$

As shown in Eq. (9), the step size δ is no longer fixed, but a function of n, expressed as $\delta(n)$. $SINR_{obs}^n$ represents the signal-to-interference and noise ratio observed at the nth time. The value of $\delta(n)$ is not only related to the current moment $SINR_{obs}^n$, but also related to the previous moment $SINR_{obs}^{n-1}$. According to the adopted adaptive algorithm, there are various calculation methods of $\delta(n)$. The simplest is Linear adjustment method.

As shown in Fig. 5, the steps of the adaptive step size adjustment algorithm (linear adjustment) are as follows:

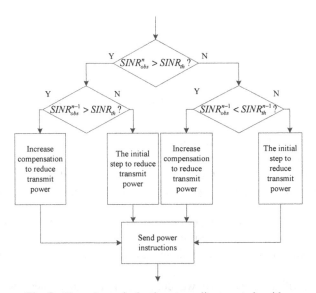

Fig. 5. Flow chart of adaptive step adjustment algorithm

(1) At time 1 (initial time), adjust the power of each aircraft with the initial step size δ_{init} according to the fixed step adjustment algorithm;

(2) At the nth moment ($n \neq 1$), the relay satellite measures $SINR_{obs}^n$ of each aircraft and compares it with the target $SINR_{th}$;

(3) If $SINR_{obs}^n > SINR_{th}$, it means that the transmitting power of the aircraft is too large, and then compare the last moment ratio $SINR_{obs}^{n-1}$ with the target $SINR_{th}$, if $SINR_{obs}^{n-1} > SINR_{th}$, it means that the current SINR is quite different from the target SINR, The adjustment step size needs to be increased on the basis of $\delta(n-1)$, that is $\delta(n) = \delta(n-1) + \delta_{init}$, otherwise, it means that the two are close, then adjust by the initial step size δ_{init}, that is $\delta(n) = \delta_{init}$;

(4) If $SINR_{obs}^n < SINR_{th}$, it means that the transmitting power of the aircraft is too large, and then compare the last moment ratio $SINR_{obs}^{n-1}$ with the target $SINR_{th}$, if $SINR_{obs}^{n-1} < SINR_{th}$, it means that the current SINR is quite different from the target SINR, The adjustment step size needs to be increased on the basis of $\delta(n-1)$, that is $\delta(n) = \delta(n-1) + \delta_{init}$, otherwise, it means that the two are close, then adjust by the initial step size δ_{init}, that is $\delta(n) = \delta_{init}$;

(5) The relay satellite generates a power control command and sends it to each aircraft. Each aircraft adjusts the transmission power according to the power control command, and waits for the system to respond to the power change before starting the next iteration.

As shown above, the adaptive step size adjustment algorithm is very flexible. When the adaptive algorithm of $\delta(n)$ is designed properly, the convergence speed will be greatly improved. However, due to the variable step size, the content of the power control instruction needs to be represented by multiple bits, which increases the system overhead.

4 Simulation Results

4.1 Simulation Results of Fixed Step Adjustment Algorithm

To compare and analyze the existing power control algorithms, it is necessary to build a simulation platform. The simulation platform is shown in Fig. 6, using the AWGN channel. The path loss calculation formula is shown in Eq. (3); in addition, thermal noise needs to be considered, which is caused by the random motion of electrons in the conductor caused by excitation [1], the calculation formula is as follows:

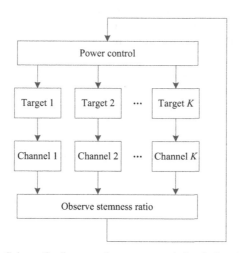

Fig. 6. Schematic diagram of power control simulation platform

$$N_R = kTB \tag{10}$$

Among them, k is Boltzmann's constant, take J/K, T represents the resistance noise temperature, the unit is K, B represents the noise bandwidth, the unit is Hz [1]. When B = 10 MHz and T = 300 K, NR = −103.8 dBm.

For the observation of SINR, the most accurate method is system simulation. Monte Carlo method is used to obtain the $E_b/N_0 - BER$ curve, but every time the power is adjusted, all the aircraft must be simulated once, which is too long and inefficient. Therefore, consider using an estimation method to evaluate the influence of multiple access interference. A common estimation method is Gaussian Approximation (GA). Assuming that the target user number is 1, the Gaussian estimation method models the multiple access interference as a Gaussian noise process. The calculation formula is as follows:

$$SINR_{GA} = \frac{1}{\frac{N_0}{E_s} + \frac{2}{3N_s} \sum_{k=2}^{K} \frac{P_k}{P_1}} \tag{11}$$

Among them, E_s represents the symbol energy, N_0 represents the unilateral noise power spectral density, N_s represents the spreading ratio, P_1 represents the power of the target user, P_k (when $k \neq 1$) represents the power of other users, that is, the source of multiple access interference. Define the power ratio $PR_k = P_k/P_1, k \neq 1$, when each user's PR_k is equal, it is simplified to PR. The spreading code is a Gold sequence with a period of 255, the spreading ratio is 255, the number of users is 50, and the modulation method is BPSK. The comparison result of the two methods is shown in Fig. 7, whether it is the case of PR = 0 dB (perfect power control) or In the case of PR = 6 dB (near-far effect), the error rate estimated by the Gaussian estimation method is very close to the actual simulation result, especially when E_b/N_0 is low, the estimation error of the Gaussian estimation method is still within the acceptable range as E_b/N_0 increases. Therefore, the simulation platform uses the Gaussian estimation method to estimate the signal-to-interference and noise ratio under multiple access interference.

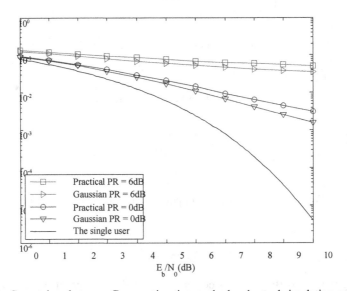

Fig. 7. Comparison between Gauss estimation method and actual simulation results

Set the simulation parameters of the power control algorithm of The multi-target TT&C system. The carrier frequency f is 2560 MHz, the symbol rate f_s is 25 kHz, the spreading ratio N_s is 255, the number of aircraft K is 50, and the position of each aircraft is fixed. The closest aircraft altitude is 200 km and the highest altitude is 2000 km, and the remaining aircrafts are evenly distributed between the above two. The initial transmission power of each user is 10 dBm, and the transceiver antennas use omnidirectional antennas. As shown in Fig. 7, when PR = 0 dB, E_b/N_0 = 8 dB, the BER is about 10^{-2}. When channel coding is used, the BER performance can meet the system requirements, so the target $\left(E_b/N_0\right)_{th}$ is set to 8 dB, referring to Eq. (7), when the absolute value of the difference between the equivalence $\left(E_b/N_0\right)_{obs}$ and $\left(E_b/N_0\right)_{th}$ estimated by the method Gaussian estimation is greater than Δ, the aircraft is considered to be interrupted.

Set Δ = 0.5 dB, the simulation result is shown in Fig. 8. When the step size is 0.5 dB, the proportion of aircraft communication interruption drops very slowly in the early stage due to the small step size. After 52 iterations, the proportion converges to 0; When it is 1 dB, the proportion of aircraft communication interruption decreases faster in the early stage. After 36 iterations, the proportion converges to 0; when the step size is 1.5 dB, the proportion of interrupted aircraft at the beginning decreases the fastest, but when the interruption ratio drops to about 40%, the interruption rate rebounded and finally stabilized at 36%. This was because the adjustment step was too long. When some aircraft were connected, the same number of aircraft was interrupted by multiple access interference. It can be seen that when the step length of the fixed step adjustment algorithm is too long, the interrupt ratio cannot converge to 0. After reducing the adjustment step size, although it can converge to 0, the convergence time is too long.

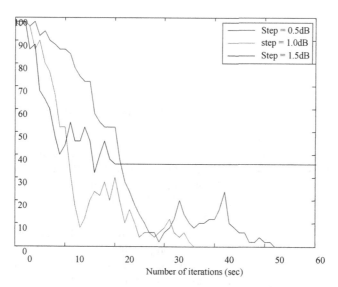

Fig. 8. Simulation results of fixed step size adjustment algorithm

4.2 Simulation Results of Adaptive Step Adjustment Algorithm

We then simulated adaptive step size adjustment algorithm, where the step size is linearly increased, and the initial step size $\delta_{init} = 0.5$ dB, the simulation result is shown in Fig. 9. After 28 iterations, the proportion of the interrupted aircraft converges to 0. Compared with Fig. 8, the fixed step adjustment requires at least 36 iterations, and the adaptive step adjustment algorithm has obvious advantages in convergence speed. During the convergence process, when iterated to the 8th and 17th times, the interruption ratio rebounded, which affected the convergence speed. Therefore, consider improving the step size adjustment algorithm to prevent rebound in the iterative process.

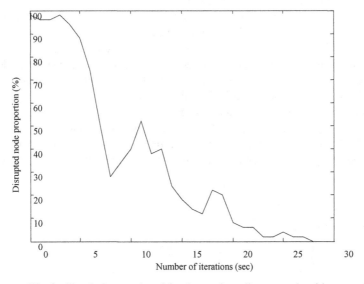

Fig. 9. Simulation results of fixed step size adjustment algorithm

The analysis shows that this rebound is caused by the increasing of the step length during the adjustment process of linear adjustment method. When the optimal solution position is close, the step size has been adjusted to a large value and the optimal solution will be exceeded in the next adjustment, then the search needs to be retraced. To solve this problem, the linear adjustment method needs to be improved. One solution is to set the upper limit of the step size, so that even if the optimal solution is crossed, the search can quickly converge to the optimal solution position again. We set $\delta_{max} = 1$ dB, the simulation result is shown in Fig. 10. During the convergence process of the improved algorithm, there is no significant rebound. After 15 iterations, it can converge to 0, which is significantly faster than the original algorithm.

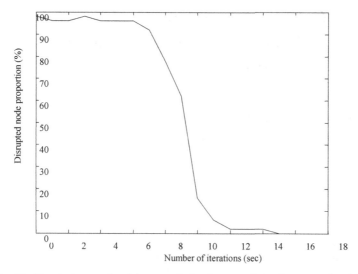

Fig. 10. Simulation results of improved adaptive step size adjustment algorithm

5   Conclusions

This paper studies the power control algorithm in the multi-object measurement and control system, analyzes two power adjustment algorithms, fixed step size adjustment algorithm and adaptive step size adjustment algorithm, and improves the adaptive step size adjustment algorithm by introducing the upper limit δ_{max} of the adjustment step size. The simulation results verify the effectiveness of the algorithm.

References

1. 啜钢, 李卫东. 移动通信原理与应用. 北京: 人民邮电出版社 (2010)
2. Tse, D., Viswanath, P.: Fundamentals of Wireless Communication. Cambridge University Press, Cambridge (2005)
3. Moshavi, S.: Multi-user detection for DS-CDMA communications. IEEE Commun. Mag. **34**(10), 124–136 (1996)
4. Verdú, S.: Minimumprobability of error for asynchronous Gaussian multiple access channels. IEEE Trans. Inf. Theory **32**(1), 85–96 (1986)
5. Bock, F., Ebstein, B.: Assignment of transmitter powers by linear programming. IEEE Trans. Electromagn. Compat. **6**(2), 36–44 (1964)
6. Aein, J.M.: Power balancing in systems employing frequency reuse. COMSAT Tech. Rev. **3**(2) (1973)
7. Zander, J.: Performance of optimum transmitter power control in cellular radio systems. IEEE Trans. Veh. Technol. **41**(1), 57–62 (1992)
8. 暴宇, 李新民. 扩频通信技术及应用. 西安: 西安电子科技大学出版社 (2011)
9. 米勒. CDMA 系统工程手册. 许希斌, 周世东, 赵明, 等译. 北京: 人民邮电出版社 (2001)
10. 刘嘉兴. 飞行器测控与信息传输技术. 北京: 国防工业出版社 (2011)

Graph Computing System and Application Based on Large-Scale Information Network

Jingbo Xu[1,2], Zhao Li[2(✉)], Weibin Zeng[2], and Jiaming Huang[2]

[1] Peking University, Beijing, China
jingbo.xu@pku.edu.cn
[2] Alibaba Group, Hangzhou, China
{lizhao.lz,qiaozi.zwb,jimmy.hjm}@alibaba-inc.com

Abstract. Graph computing is more and more widely used in various fields such as spatial information network and social network. However, the existing graph computing systems have some problems like complex programming and steep learning curve. This paper introduces GRAPE, a distributed large-scale GRAPh Engine, which has the unique features of solid theoretical guarantee, ease of use, auto-parallelization and high performance. The paper also introduces several typical scenarios of graph computing, including entity resolution, link prediction, community detection and graph mining of spatial information network. In these scenarios, various problems have been encountered in the existing systems, such as failure to compute over large-scale data due to the high computation complexity, loss of accuracy due to the cropping of original data and too long execution time. In the face of these challenges, GRAPE is easy to support these computing scenarios with a series of technical improvements. With the deployment of GRAPE in Alibaba, both effectiveness and efficiency of graph computing have been greatly improved.

Keywords: Graph computing · Distributed system · Incremental evaluation · Information network

1 Introduction

The need for graph computations is evident in a multitude of use cases. Graph is widely used since it is nature to model entities and represent the relationships among them with graph. In the scenario of big data today, spatial information network, social network, knowledge graph, transportation and communication network, supply chain and logistics planning are all typical examples of graphs(as shown in Fig. 1).

Study and application related to graph data have been a focus in the academic and industrial circles, which can be classified into different directions according to different emphases. They mainly includes:

© Springer Nature Singapore Pte Ltd. 2021
Q. Yu (Ed.): SINC 2020, CCIS 1353, pp. 158–178, 2021.
https://doi.org/10.1007/978-981-16-1967-0_12

- Graph query/travel, such as Neo4j [33], JanusGraph [10], Amazon Neptune [26] and GraphCompute of Alibaba Cloud [25], which provide the management of graph data and OLTP-oriented real-time query;
- Graph computing, such as Powergraph [20], Pregel [31] and GraphX [21], which focus on more complex OLAP-based offline analysis;
- Other directions, such as knowledge graph, graph embedding, graph data quality and query language design.

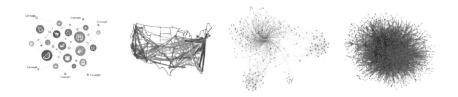

Fig. 1. Graphs on the Internet.

Among these directions, graph computing, which serves as an effective tool to analyze and dig the mass associated data, is widely used in many scenarios. The core of graph computing lies in the algorithms. The common graph algorithms and their application include: (1) Simple graph analysis: such as PageRank, shortest path, graph coloring, connected component and reachability; (2) Community detection: triangle counting, which can be used to analyze the common neighbor relationships of the nodes; K-core analysis can look up the largest subgraph, in which the degree of each node is larger than k. K-Truss analysis can look up a subgraph, in which each edge should be included in at least k-2 triangles; such algorithm can be applied in graph structure extraction, visualization, vertex structure analysis and so on; (3) Pattern matching: graph simulation, graph isomorphism and keyword search; (4) Collaborative recommendation: alternating least squares, stochastic gradient descent (SGD) and tensor factorization; (5) Structure prediction: such as loopy belief propagation, max-product linear programs and Gibbs sampling.

Based on the above common algorithms, graph computing can be applied to the following scenarios. First of all, graph computing can be used for social marketing. Nelson, an American famous consulting agency, pointed out in its marketing research report on social media that "90% consumers tend to trust the mutual recommendations on social networks, while only 14% still trust advertisements" and "social media marketing will completely take place of traditional marketing patterns" [34]. This has also been proved by the marketing capacity of China's sharing platforms like Xiaohongshu and Weibo. In the recommendation systems, recommendations based on association rules, such as collaborative filtering and frequent itemset mining, have been proved to be an effective way. The introduction of social networks in the form of graph data into association rules will offer more possibilities for precise recommendations. With the help of graph data, we can describe the target groups better, and thus translate relevant recommendations into graph pattern matching [16]. For example, Fig. 2 represents some association rules based on graph pattern. Another example is a public crisis faced by Facebook in recent years: Cambridge Analytics, a British data analysis company, conducted analysis on partial data of Facebook and accurately delivered suggestive election advertisements to users, thus affecting the trend of the US election to a certain extent [22]. The outbreak of this incident, on the one hand, exposed the hidden concerns of the public's privacy security in the era of big data, and on the other hand, showed the accuracy of social recommendations based on graph data from a technical perspective.

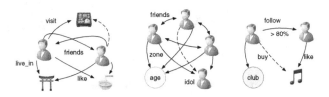

Fig. 2. Association rules based on graph pattern.

Another field in which graph computing can exert its unique algorithm advantage is financial risk control. In the context of the rapid growth of China's consumer finance, a number of companies that offer rapid lending have emerged, which allow users to apply via mobile phone and provide them a safe lending quota through risk assessment. During this process, financial fraud is the major risk they are faced with. With the help of graph data, we can establish a common feature and spatial-temporal information network among customers based on information such as user location, consumption time, mobile phone number, company, address, bank account, social security record, transaction order form and so on, as shown in Fig. 3. The customer nodes in the figure include some known high-quality customers, some customers with unknown credit, as well as a very small number of high-risk customers and known fraudulent customers. Through relevant algorithms of graph computing, such as label propagation, shortest path, reachability, density monitoring and spatial graph, we can solve a series of problems related to risk control and credit investigation.

Graph computing is also of great significance to AI. At present, machine learning has been widely used in big data analysis and AI, with fruitful results being achieved. On the one hand, machine learning and graph computing complement each other in the field of AI. Machine learning focuses on dense matrices, such as image processing; while graph computing is more suitable for sparse matrix processing. With the introduction of concepts like node2vec, graph embedding serves as a bridge between graph computing and machine learning by providing a series of methods to map sparse data into a dense space, so as to apply the traditional machine learning methods. In recent years, with the rise of graph neural network, graph has become even more inseparable from machine

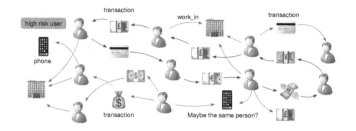

Fig. 3. Money laundering loop in financial risk control.

learning. Alibaba also has some open-source graph-based machine learning tools, such as [24,43], which have been verified by practical application scenarios. On the other hand, theoretical limitation of machine learning is an undeniable fact. For example, Turing Award winner Judea Pearl believed that the existing machine learning systems operated almost as statistical or blind models, and the causal logic behind the results was a black box, which limited the further development of machine learning [35]. One possible solution to it is graph-based structural causal reasoning. Currently, the interpretability of machine learning has begun to get attention from academic and industrial circles. Some studies have begun to explore the logical relationship between the input and output of machine learning based on graph data structure, trying to reveal the reasoning logics of machine learning in natural language processing, semantic model analysis and other fields, so as to enable researchers to conduct optimization and further improve the performance of machine learning and expand its application scope.

The remainder of the paper is organized as follows: Sect. 2 introduces the research work related to graph computing engines, Sect. 3 introduces the design and characteristics of the graph computing system GRAPE, Sect. 4 introduces some typical scenarios of graph computing in Alibaba, Sect. 5 introduces the problems encountered in the application of the existing graph systems in these scenarios, Sect. 6 introduces how GRAPE system addresses these practical problems through technical optimizations, Sect. 7 introduces the effect of GRAPE in addressing these problems, and Sect. 8 makes a summary of the whole paper and puts forward the outlook on future work.

2 Related Work

The huge amount of data available, low-cost storage and the success of online social networks have led to the explosive growth of data size. In this context, the graph data that systems need to process has become quite massive. For example, the data of the largest publicly available graph in open data set clueweb12 [13] is 120 GB, which contains 1.4 billion nodes and 6.6 billion edges; inside Alibaba, it is very common that the data size of a graph-based model is tens of billions or even hundreds of billions of edges. According to Literature [15], the graph data of Facebook is roughly at the scale of one trillion.

With such a large scale, the traditional single-machine serial graph algorithms have become incapable of efficient execution. Some studies, such as [32], have questioned the efficiency of distributed computing, believing that a high-performance single computer can solve the problem of graph computing. Nevertheless, the types of graph data and computing scenarios are far more complicated than they were assumed in the experiment in [32]. And considering the fault tolerance, single-machine cost and system complexity, distributed systems are more favored in the industrial circles. In this context, a series of distributed graph computing systems have developed rapidly in recent years. In a paper published in 2010, Google introduced its internal large-scale distributed graph computing system Pregel for the first time [31]. Pregel is a parallel graph processing system implemented [40] based on a synchronous parallel model that uses a vertex-centric programming model. Under this model, each vertex on the graph has a same computing task, which only reads the data stored in its own and neighbor nodes for processing. The whole computing process is composed of a series of supersteps (a superstep refers to one iteration in the calculation), each node waits for one global synchronization after calculating one superstep, and the iteration state will last until a certain termination condition of the algorithm is reached. Pregel is known as one of the major parts of Google's new "Troika" in the post-Hadoop era that has influenced the development trend of big data once again. Apache Giraph [11] is realized based on the open source of this model. In addition, a series of graph computing platform, such as GPS [38], Pregel+ [9] and Mizan [29], as well as Alibaba's ODPS Graph, have all inherited the Pregel model (Fig. 4).

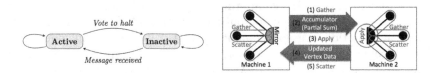

Fig. 4. Models of Pregel and PowerGraph, based on [31] and [20].

Literature [20] presented PowerGraph in 2012, a large-scale graph computing engine designed for machine learning. PowerGraph proposed another model: Gather-Apply-Scatter (GAS). The GAS model also take the vertex as the computing center and operates in the mode of collecting data from neighbor points, and processing, updating and triggering the neighbor points to conduct the next round of processing. It breaks the shackles of synchronization in the BSP mode and makes PowerGraph process power-law graph more efficiently with the help of point segmentation strategy. The GAS model also has many followers, such as Spark GraphX [21], PowerSwitch [8] and PowerLyra [14]. PowerGraph was renamed Turi and was purchased by Apple in 2016, and later it was made an open-source project known as TuriCreate [3]. As a link in Apple's AI ecology, it is used for developing machine learning models. This has also indirectly confirmed the prospects of graph computing in the field of AI.

Though these graph computing engines can solve the parallelization problem of large-scale graph computing, they usually only support the programming methods that are based on vertex-centric models. Some other studies Yan et al. [42], Tian et al. [39] proposed the structures based on subgraphs or blocks, but they still inherited the programming methods of vertex-centric models in terms of the granularity of superpoint or subgraph. Vertex-centric programming requires users to have a deep understanding of relevant fields, making graph computing a patent held by only a few algorithm experts. And on the other hand, studies on graph computing have lasted for several decades and a large number of algorithms have been accumulated, but these algorithms are often serialized and need to be parallelized by distributed experts before being executed on these platforms.

3   GRAPE Framework

How to maximize the reuse of the existing graph computing algorithms? How to improve the programmability so as to lower the threshold of graph computing? How to guarantee the correctness and termination of parallel graph computing? To solve these problems, we developed GRAPE (GRAPh computation Engine) [17], a large-scale distributed graph computing engine that adopts a brand new model. The architecture design of GRAPE is as shown in Fig. 5.

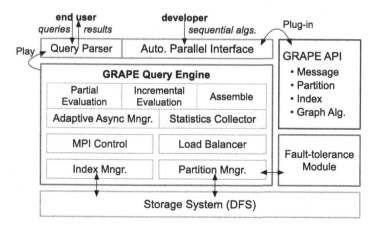

Fig. 5. GRAPE architecture.

In the architecture of GRAPE, static graph data is stored as a distributed file system in the bottom layer. The core engine layer is in charge of processing partial evaluation, incremental evaluation and result assembling, which is composed of a series of components, including message controller, index management module, partitioned management module, load balancing module and error processing module, etc. GRAPE is designed for two kinds of users: end users only have the demand to use graph computing and they can directly invoke the algorithms in GRAPE's built-in algorithm library for graph computing; senior users can develop custom graph computing programs through GRAPE's programming interface and realize the plug and play of algorithms in GRAPE.

Compared with the existing graph computing engines, GRAPE has the following core advantages:

1. Solid theoretical guarantee: GRAPE extends from the fixed point computing theory of partial evaluation/incremental evaluation. In a distributed environment, each compute node conducts partial evaluation in the initial round according to the data it holds, and then through several rounds of information exchange, where each round takes the information from other compute nodes as updates, incremental evaluation is triggered, and the calculation will end when the system gets stable. This process is proved by rigorous theories [27], and GRAPE is terminative and accurate as long as the preconditions are satisfied.

2. Programmability: in addition to supporting Pregel model, GRAPE has also proposed a subgraph-based calculation model called PIE (PEval-IncEval-AssEmble), under which users need to provide three functions:
   - A *PEval* function used for partial evaluation, it is usually a single-machine sequential algorithm;
   - A *IncEval* function, a sequential incremental algorithm, used for incremental evaluation;
   - function *Assemble* used for result assembling, such as summing the results. Among the above three functions, the single-machine sequential algorithms used for partial evaluation and incremental evaluation are already exist. Users can also write codes according to the single-machine graph computing logic when necessary. Compared with Pregel model, PIE model avoids users from redesigning and rewriting the graph computing programs based on a vertex-centric model, providing a more intuitive and simple programming logic.

3. Automatic parallelization: when a user provides three functions, GRAPE will compile the algorithm and distributes it to each executing work node (worker). Each node will first invoke partial evaluation function PEval, and this computing state will be preserved and a message will be generated when a boundary point is encountered. This message will be received and processed by other workers in the next round of calculation. After a worker receives a message, it will merge the message according to a user-specified message conflict resolution function, and eventually act as an update to trigger the current round of incremental evaluation. Calculation will end when no new message is generated by any worker.

4. Performance advantage: in terms of open data sets, GRAPE has a significant performance advantage over the existing open source systems.

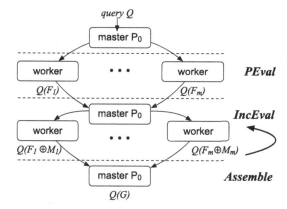

Fig. 6. Workflow of GRAPE.

Combining with GRAPE's workflow in Fig. 6, let's explain in a more detailed manner how GRAPE works and achieves the above effects. Assuming that a graph named G is partitioned by data into n subgraphs, namely $(F_1, F_2, \ldots F_n)$, which are distributed at n workers $(P_1, P_2, \ldots P_n)$ in a distributed environment. When a query $Q \in \mathbb{Q}$ and a partitioned graph G is input, GRAPE will calculate $Q(G)$ according to the following steps:

1. Given a function $f(s, d)$, where s is a partial input, and partial evaluation is done based on the known partial input s [37]. It will calculate the part related to s, and then translate f into function f' that only relates to unknown input d. At each compute node P_i in GRAPE, graph data partition F_i it holds is the known input s, and the partitions at other compute nodes are regarded as unknown input d. GRAPE will trigger $Q(F_i)$ at each node in parallel as partial evaluation.
2. Incremental evaluation. GRAPE will exchange the result of part of partial evaluation between nodes in the form of messages, and take message M_i sent to P_i as the increment of the graph to calculate $Q(F_i \oplus M_i)$. It should be noted that message M_i here carries some edge points/edges on partition Fi as well as the variables on them. What GRAPE does in this process is incrementally "adding" part of the new computed result O_i to the original result $Q(F_i)$, such that $Q(F_i \oplus M_i) = Q(F_i) \oplus O_i$. In general scenarios, M_i and O_i are relatively small, making incremental evaluation more efficient than recalculation of $Q(F_i \oplus M_i)$. Another advantage is that incremental evaluation is often bounded: its overhead is only related to message size M_i and the variable quantity of result set O_i, while having nothing to do with the whole data size F_i [37].
3. When incremental evaluation reaches a fixed point, the algorithm will converge and no new computation will be conducted. At this moment, worker P_0, as the master compute node, will draw the partial results from each

node, and invoke a user-specified simple aggregation function Assemble to aggregate the results and output them.

Notations mentioned in the above steps are shown in Table 1.

Table 1. Notations

Notation	Description
$Q, \mathbb{Q}$	A graph query class and a graph query, where $Q \in \mathbb{Q}$
G	A piece of graph data, either directed or undirected
P_0, P_i	P_0: master compute node, P_i: other compute nodes
F_i	A partition of data graph
M_i	Message designated for compute node $P_s i$
O_i	Partial result generated on $P_s i$

4   Applications of Graph Computing

In order to verify the theoretical model and practical application ability of GRAPE, we support some computing tasks based on real business scenarios and realistic demand. This chapter introduces several graph computing scenarios.

4.1   Entity Resolution

The Internet is profoundly affecting the lifestyles of users, and the convenience of life provided to users is becoming more and more diversified. Users can call for a takeout, watch movies, purchase household appliances and even complete a new house decoration through the operation on mobile phones or web pages. The data about shopping, VOD (video on demand) and tourism are scattered in various services, and users may use these services in different websites and terminals. Which actions are initiated by the same user? Which product descriptions on different service providers actually correspond to the same item? How to find the real-world entity mapping under the premise of ensuring privacy is a type of entity resolution problem that academia and industry want to solve. The solution to this problem is explored in [12]. Entity resolution can integrate the feature information of different data sources to provide an accurate and personalized portrait basis for scenarios such as advertisement recommendations.

Alibaba's current ecological construction includes the core e-commerce business, as well as the entertainment, local life and finance sectors. Each business format is a data source. In order to make these data drive the business and play a more valuable role, Alibaba has built a global data OneData system [1]. In the OneData system, through a series of processes such as data access, specification definition, calculation processing, data verification and data stability, data silos are eliminated, data are connected, data accommodation is realized, and the accuracy and comprehensiveness of data applications such as business analysis and user portraits are ensured. In the process of OneData global data construction, a critical step is entity resolution. The data need to be clustered through the Connect Components (CC) in graph computing, that is, the vertices in the connected components are considered to belong to the same entity.

4.2 Social Relationship Prediction

In the context of the Internet, a social network is a large-scale graph, which often changes the network structure dynamically over time. How a social network graph will evolve in the future is a classic graph computing problem - link prediction. Link prediction refers to predicting the possibility of a connection between two nodes that have not yet connected with the known network structure in the graph [19]. There are many application scenarios for link prediction on social networks, such as recommending possible friends to users, team recommendations for various games or activities, etc.

In the relationship analysis and link prediction on social networks, some features will be used to measure the possibility of an edge between two nodes in the future. Among them, Common Neighbors (CN) is an important feature. It marks that two points that are not directly related have a common connection relationship with the third point. In a social network, it is expressed as a common friend relationship between two users who are temporarily not friends with each other. In addition to common neighbors, there are some other features, such as spatial information between nodes, Jaccard coefficients, Adamic-Adar (AA) coefficients, Resource Allocation (RA) coefficients, etc. [19]. These features characterize the structure of the graph in various dimensions and provide key information in the heuristic prediction of graph links.

4.3 Community Detection

Community detection, as one of the classic problems in graph theory, has attracted wide attention for a long time. There are a variety of algorithms for the community detection [28]. In terms of classification, there are betweenness-based partitioning algorithms such as the GN algorithm, modularity-based algorithms such as FN algorithm and Louvain algorithm, and dynamic algorithms such as LPA algorithm and infomap algorithm based on random walk.

The application of community detection is very extensive. For example, in the community detection of interest forums, user groups with different interests and backgrounds can be discovered to match different promotion strategies; In the transaction network, user groups with different purchasing preferences and price sensitivity can be classified, and advertisements can be placed according to group characteristics; In payment and financial networks, groups that maliciously defraud concessions and engage in click farming can be identified.

As shown in Fig. 7, in finance-related businesses, anti-money laundering and anti-fraud have always been an important task to ensure financial security.

Due to the limited resources of the dark industry, there is a natural aggregation and correlation between cheating accounts, and a close community will eventually be formed. Therefore, the role of risky community detection in finance and other related businesses is crucial. In addition, due to the strong correlation between the dark industry accounts, we can build a relationship network between accounts based on the correlation between the accounts and conduct risk communication in the network based on the identified dark industry accounts, so as to dig out more cheating accounts and gangs engaged in dark industry. This is a typical application of multi-labels propagation (MLPA) in community detection.

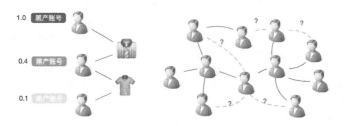

Fig. 7. Community detection of MLPA and link prediction in social networks.

4.4   Graph Mining of Spatial Information Network

The relationship between nodes in a spatial information network is determined by the physical distance between entities. Through the construction of a spatial information network, the complex spatial information relationship between entities can be effectively modeled. In the e-commerce risk control scenario, multiple devices controlled by the same dark industry are closely related in space. Therefore, a spatial information network between devices can be constructed based on the spatial information and behavior data between the devices. Spatial information network is crucial to the e-commerce risk control.

In the process of spatial information network mining, an effective method is graph embedding. The literature [36] started the research boom of Graph Embedding. Algorithms represented by Deepwalk, Node2Vec [23], Struct2Vec [18] and GraphSage [30] have received widespread attention in the industry, and their related applications emerge one after another incessantly. Moreover, the researchers used the ideas of convolutional networks, recurrent networks and deep autoencoders to define and design neural network structures for processing graph data, which gave rise to a new research hotspot - Graph Neural Network (GNN) [41]. There are many categories of graph neural networks, such as Graph Convolution Networks (GCN), Graph Attention Networks, Graph Autoencoders, Graph Generative Networks, and Graph Spatial-temporal Networks [41].

With the rise of deep learning and GNN, the method of using graph data as the model and deep learning as the core has gradually become the mainstream method of spatial information network mining. In the e-commerce risk control scenario, real-time risk identification, prevention and control are required. Therefore, the graph mining system based on the spatial information network needs to realize the ability of real-time computing.

5   Existing Problems

The graph computing scenarios introduced above have been applied in multiple scenarios such as e-commerce recommendations, social networks, logistics systems, and financial risk control. However, the existing technical solutions have disadvantages such as difficulty in large-scale application, low timeliness and high implementation complexity. For example, the following difficulties and challenges exist in Alibaba's existing systems:

- Many business is deployed in the big data computing service MaxCompute [5]. However, due to the limited resources of the computing resource pool in MaxCompute, when the scale of graph computing is large, considerable resources are required. For example, some entity resolution tasks may require more than thousands of worker nodes, and it takes too long to wait for the resources to be ready on the MaxCompute platform.
- When the task is large and the number of worker nodes rises, another negative impact is the generation of stragglers: Due to the uneven job splitting, the end time of a task is slowed down by one or two long-tail worker nodes. To make matters worse, sometimes a task requires thousands of worker nodes to calculate at the same time. If a lagging worker node finally fails after being executed for a long time, the entire calculation task would fail and the calculation task needs to be restarted.
- Some algorithms have simple logic and are easy to implement on a single machine, but as the scale increases, the complexity of the problem increases exponentially, such as the algorithm for finding common neighbors (CN). Although graph modeling is intuitive and the algorithm is easy to implement, in practice, limited by the data scale of the graph and the graph processing capabilities of MaxCompute, graph computing cannot be used directly. In fact, because the scale of the social network graph to be processed is at the level of tens of billions, MaxCompute's graph module cannot load the data. Therefore, the business team previously used MaxCompute's SQL with custom function calls for computing. Moreover, in order to make the process run successfully as much as possible, the business team introduced some lossy approximations and data size trimming at the algorithm design level. This kind of approximation and trimming has a potential impact on the accuracy of the algorithm and may cause errors or omissions in the hidden value of the original graph.

In the risk control scenario, it is necessary to perform spatial information graph mining computing based on the user's historical behavior records and spatial data. Affected by the performance of MaxCompute on graph data, the

SQL + Tensorflow solution is currently adopted. However, the problem of this solution is that it takes a long time to find the association relationship on the graph through the time-consuming join operation in SQL. Moreover, since label diffusion requires many rounds of iterations, it is necessary to wait for resource allocation and readiness between iterations. Since the time for one computing is too long, it cannot well meet the dynamic offensive and defensive needs of high timeliness and strong game in the risk control scenario.

6 GRAPE-Based Solution Design and Implementation

The stand-alone algorithms of the business team are all basic graph algorithms. How to apply to large-scale graph data and how to parallelize such algorithms in a distributed manner are the problems that GRAPE is committed to solving. At the same time, in the process of business implementation, some problems that were ignored in the design of GRAPE prototype at the conceptual level were observed, and some challenges were encountered. In this chapter, we will introduce the design and implementation of the solution based on GRAPE, as well as some technical points.

6.1 Optimization of End-to-End Time

For the first application entity resolution, CC is a built-in algorithm package of GRAPE, which has excellent performance [17]. For example, it only takes 8 s for the GRAPE prototype system to perform the task once, while the original computing time needs at least half an hour. However, these data only represent the core computing performance, and there are still some problems when it is actually deployed and delivered to the business team. For example:

- The prototype of GRAPE supports multiple computations for one-time loading of graphs: After loading graph data once, users can load multiple algorithms, such as CC, PageRank and reachability/shortest path calculation, and support multiple independent queries until there is no more computing task for this graph. In the typical analysis task of the business team, the loading of a graph only corresponds to a computing, which exposes the problem that the end-to-end time of the prototype is too long.
- Since the main focus of GRAPE is put on the computing performance in the prototype system, the input/output (I/O) is not optimized. Although it supports reading and writing data from local files, HDFS, Amazon S3 and other locations, the full graph data at each computing node will be scanned and read, and there is no distributed segmented reading and writing and multi-threaded operations. As a result, the proportion of the I/O time in the end-to-end time will be too high.
- The GRAPE prototype either partitions the graph before computing or directly loads a pre-partitioned file. However, when the task is only run once, the benefit of partitioning the graph once after loading the graph is limited. If the user is required to provide a pre-partitioned file, it is unrealistic for most users as their data are not even stored as a graph in the data warehouse.

In response to the above problems, we have added some new support to GRAPE to optimize end-to-end task time and business results:

- Support variable-length character strings as the point identifier of the graph data, without the need for the business team to convert the data format, and directly read and write business tables from upper and lower operations.
- The OSS read-write adapter is developed based on the OSS SDK [7], which supports the efficient reading and writing of data from the OSS Bucket or the OSS-based ODPS external table. Multiple computing nodes of GRAPE can each pull part of the files in the Bucket, and then exchange data separately.
- Based on ODPS Tunnel [6], an ODPS read-write adapter that supports segmented reading is developed, which supports efficient multi-threaded reading and writing of ODPS data tables. GRAPE will pull data in multiple threads within each computing node, and at the same time, a thread will process the pulled data and start building the graph. The network delay and the time cost of building the graph are overlapped to the maximum extent.
- A Hash-based partitioning module for streaming data is developed, so that pre-partitioned data or fine partitioning is no longer required.

Through these methods, we have effectively improved the end-to-end time of GRAPE, delivered the entity resolution algorithm to the business team.

6.2   Optimization of Memory Usage

GRAPE encountered greater challenges when dealing with link prediction tasks. First, the scale of graph data is large. The test data graph of this task has about 30 billion edges. It takes about 4 h to load the graph. After loading, it occupies more than half of the total memory of the cluster machines. Secondly, unlike CC that only changes the ComponentID on the point, the common neighbor (CN) algorithm will calculate and output each dotted pair and the common neighbor value between them, during which massive amounts of intermediate result data will be generated. In the preliminary test, the calculation failed since the memory was fully occupied. In a real social network graph, the amount of data on the side can even reach hundreds of billions.

Through the analysis of the CN algorithm, we found that some internal points might no longer participate in the subsequent process after the current stage. These "intermediate results" are actually the final results. Based on this, we added a mechanism for writing results in advance for GRAPE, so that GRAPE could write the results during computing. In addition, we introduced some auxiliary structures to determine whether a point had been calculated, and if it was, then it would be output as soon as possible. In this way, we largely reduced the memory occupation. This mini-batch method is a universal abstraction. We provide it as an optional operating mechanism of the GRAPE engine, and it is also applicable in some other scenarios, such as graph sampling.

At the same time, we optimized the memory management of GRAPE. For example, GRAPE uses a compact vertex index structure internally for efficient computing. During data input and result output, a VertexMap needs to be established and the point ID is converted according to this VertexMap. We optimized this structure, so that it could flexibly support sharing among multiple computing processes on the same machine, mapping to hard disk for reading and writing, or distributed maintenance of computing nodes. After a series of optimizations, the CN task was successfully executed. Due to the migration to the graph as the basic structure of the computing and the removal of some approximate methods in the UDF, the computed results are more abundant than the original solution. In addition, we extended the process according to the needs of the business team. While calculating the CN, we calculated two additional core features of RA/AA [28], which improved the accuracy of link prediction.

6.3 Operations

Providing stable services to the business team is the basic requirement for GRAPE to take over the production process online. The GRAPE prototype system only contains core modules. Based on Alibaba's powerful one-stop big data intelligent cloud research and development platform - DataWorks [4], we deployed GRAPE services on it and accessed to the user's task line. The deployment structure is shown in Fig. 8 The docking work includes the following transformation and adaptation parts:

- Deployed the GRAPE high-performance computing resource pool and managed it through k8s;
- Developed GRAPE Service based on node.js, providing RESTful call interface for GRAPE tasks;
- Developed the corresponding configuration node on Dataworks, so that users could configure task parameters through the form;
- Deployed fault tolerance and automatic retry mechanism to ensure the normal execution of tasks as far as possible;
- Access to the unified authorization and alarm mechanism of DataWorks.

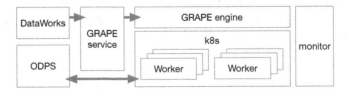

Fig. 8. Deployment of GRAPE.

6.4   Real-Time Update Support

In the demand for real-time community detection, the graph to be calculated is dynamically updated through streaming data. In this case, the community detection algorithm should be able to be queried in the latest data in real time as much as possible. The GRAPE prototype system supports multiple computations for one-time loading of graphs. On this basis, we have implemented an additional graph update component that can change the graph topology and data already loaded into the memory. Based on flexibility and performance considerations, we provide mutual conversion between variable and immutable graphs in memory. At the same time, Apache Kafka [2] is introduced to solve the problems of fault tolerance and consistency and ensure the final consistency. The real-time system is shown in Fig. 9. The procedures are as follows:

- After the preorder steps are completed every morning, the applicable party initiates a task request through the RESTful interface, and GRAPE starts and loads the offline graph data as the base data graph;
- After the initial upload of the graph is completed, GRAPE obtains real-time data from the Kafka message queue and appends them to the graph; GRAPE can catch up with the real-time data stream soon;
- According to different computing tasks, GRAPE supports two different ways to initiate queries: 1) Quasi real-time query: In some offline computing scenarios, a query request will be initiated every hour, and the request will enter the Kafka message queue; 2) Real-time query: In some real-time computing scenarios, the query may intensively enter the Kafka queue as the graph updates.
- After GRAPE obtains the query request in the Kafka queue, it will suspend the update of the graph and perform the query on the snapshot of the current graph data. After returning to the result, it continues to get data from Kafka to update the graph.

Based on the above solution, the real-time detection capability of the community can be realized, thereby greatly improving the timeliness of entity resolution in business scenarios.

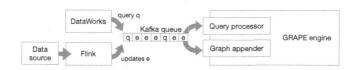

Fig. 9. Deployment of GRAPE with real-time update support.

7   Evaluation

GRAPE has now replaced the original solution to support these business online, and achieved good performance and effect. In this chapter, we report some performance data and its effects. All experiments are conducted on Alibaba's internal platform, in which the original solution is used as the baseline data.

7.1   Entity Resolution

The main graph computing link is the computing of Connected Component (CC). The data used for testing come from a daily general task, in which the graph size is about 1.2 billion edges. The end-to-end task time of the process has reduced to about 8 min from the original about 40 min to 2 h, in which the computing time is only about 90 s.

7.2   Social Relationship Prediction

The main calculation link of the social relationship prediction process is feature calculation (CN/RA/AA). The scale of online tasks is at the level of tens of billions of edges, and the data used in the experiment are a graph with about 90 billion edges. Compared with the original solution that is unable to use graphs for calculation, GRAPE directly uses graphs and graph algorithms and does not have to take approximation; The end-to-end time has been reduced from approximately 7 h obtained through approximate calculation after the data were trimmed to approximately 3.5 h, of which the calculation time is approximately 2.5 h. It should be noted that when GRAPE is used for calculation, the data are not trimmed and the approximate algorithm is not used, so it provides richer results than the original solution: While CN is calculated, two additional features (RA and AA) are calculated, which improves the core index (F1 Score) of the entire process.

7.3   Community Detection

In the application of community detection, we use multi-labels propagation (MLPA) as the test scenario. On several categories of data in a single day, the average time of one iteration reduced from 600 s to 22 s; On the 30-day data of all categories, the average time of one iteration reduced from more than 1 h to about 58 s; An online task usually contains more than 10 rounds of iterative calculations, and the calculation time has been reduced from more than 10 h to about 600 s;

Figure 10 shows the effect of the above three experimental data.

7.4   Spatial Information Network Graph Mining

In the e-commerce risk control task, an important scenario is the risk graph mining based on the user spatial information network. In practice, we build a

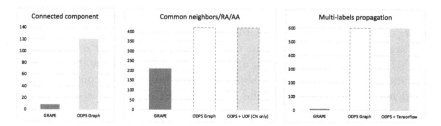

Fig. 10. Performance evaluation.

spatial information network between accounts based on the account's spatial information and account behavior records in the e-commerce scenario. The node types of the graph mainly include users, commodities and cities. The relationship of the graph is mainly based on the interactive behavior between users and commodities. The graph contains hundreds of millions of nodes and billions of edges. The overall process includes three parts, i.e., graph construction, sampling and graph computing. In the graph sampling stage, we propose spatial information sampling operators in complex heterogeneous graphs. Specifically, we first calculate the weight of each spatial node (city) as follows:

$$w_c = \frac{log(D(c))}{\sum_i^{|C|} log(D_i)}$$

Where c represents a specific city node, $D(c)$ represents the degree of the current city node, and $|C|$ represents the total number of cities in the graph. Smoothing by the log operator can prevent sampling bias caused by popular cities.

Then in the sampling stage, a hierarchical sampling method based on spatial information is introduced. We first sample the spatial information node corresponding to the user based on the weight of the city node and then sample the neighbor commodities of the corresponding spatial information node. The hierarchical sampling algorithm is adopted to avoid sampling performance problems caused by some hot commodities. At the same time, the weight of the spatial node is introduced to improve the effectiveness of sampling, thereby improving the overall performance of the model. Finally, algorithm innovation improved the stability and accuracy of the system. The overall algorithm AUC increased by 3.82% to 80.18%, and F1-score increased by 5.88% to 76.32%.

In the system implementation part, taking into account the requirements for timeliness and accuracy in the risk control scenario, we adopted the strategy of combining offline graph uploading and real-time update to compose the graph. Specifically, we first build an offline graph based on account historical space information and behavior records, and preload the offline graph into the GRAPE system. Then through Kafka, the real-time data of the day is sent to the GRAPE system for dynamic graph update. The update operations include node, edge addition and edge weight update. At the same time, we sample the binary-hop neighbors of edges and nodes in real time for graph mining computing tasks.

Practical application shows that the efficiency of real-time graph mining is very high. The peak value of graph update sampling is about 400,000 times per second, and the average speed is about 200,000 times per second. The overall system delay is within seconds, and the detection efficiency is 56.25% higher than the existing methods.

8   Conclusion and Outlook

This paper introduces the system design and business application and practice of large-scale distributed graph computing. In response to the complex graph computing programming model and high user threshold, we have developed a new generation of graph computing system - GRAPE. GRAPE is based on the theory of fixed point calculation, with theoretical guarantees for convergence and termination. At the same time, the programming model based on subgraphs is more in line with users' habits. Users can reuse existing stand-alone sequential algorithms and achieve PNP (Plug-and-Play). GRAPE can automatically parallelize algorithms to run on distributed environments and large-scale graph data. This paper also introduced the classic graph computing scenarios, as well as the main challenges encountered in practice. Through a series of technical optimization and adaptation, the GRAPE system has maturely supported these graph computing scenarios. It has been widely used in the Alibaba system and has achieved significant improvements in performance and effectiveness.

In the future work, we will continue to carry out research work in the following directions. First, we will explore more efficient partitioning algorithms, such as using specific graph algorithms for specific partitioning [4]. Secondly, GRAPE requires the user to provide an incremental version of the stand-alone sequential algorithm. For some non-classical problems, it is difficult for users to develop an incremental version by themselves. In this context, it is also our follow-up research subject to explore the automatic incrementalization of a stand-alone algorithm. Finally, GRAPE will actively embrace the Python ecology and open up the current Python-side rich toolkits and algorithm packages, such as Numpy/PyTorch/Tenserflow, to provide a one-stop platform for data scientists to do graph computing and data analysis and processing on large-scale data.

References

1. Alidata (2017). https://dt.alibaba.com/page21.htm
2. Apache Kafka (2017). https://kafka.apache.org/
3. Turi Create: simplifies the development of custom machine learning models (2019). https://github.com/apple/turicreate
4. Aliyun dataworks (2020). https://help.aliyun.com/product/72772.html
5. Aliyun maxcompute (2020). https://help.aliyun.com/document_detail/27800.html

6. Aliyun odps (2020). https://cn.aliyun.com/product/odps
7. Aliyun oss (2020). https://www.aliyun.com/product/oss
8. PowerSwitch: Adaptive Prediction and Mode Switch on Graph-parallel Computation (2020). https://ipads.se.sjtu.edu.cn/projects/powerswitch.html
9. Pregel+ (2020). http://www.cse.cuhk.edu.hk/pregelplus/
10. Authors, J.: JanusGraph (2017). https://janusgraph.org
11. Avery, C.: Giraph: large-scale graph processing infrastructure on Hadoop. In: Proceedings of the Hadoop Summit, Santa Clara, vol. 11, vo. 3, pp. 5–9 (2011)
12. Brizan, D.G., Tansel, A.U.: A survey of entity resolution and record linkage methodologies. Commun. IIMA **6**(3), 5 (2006)
13. Callan, J., et al.: The clueweb12 dataset (2013)
14. Chen, R., Shi, J., Chen, Y., Zang, B., Guan, H., Chen, H.: PowerLyra: differentiated graph computation and partitioning on skewed graphs. ACM Trans. Parallel Comput. (TOPC) **5**(3), 1–39 (2019)
15. Ching, A., Edunov, S., Kabiljo, M., Logothetis, D., Muthukrishnan, S.: One trillion edges: graph processing at Facebook-scale. Proc. VLDB Endow. **8**(12), 1804–1815 (2015)
16. Fan, W., Wang, X., Wu, Y., Xu, J.: Association rules with graph patterns. Proc. VLDB Endow. **8**(12), 1502–1513 (2015)
17. Fan, W., Xu, J., Wu, Y., Yu, W., Jiang, J.: GRAPE: parallelizing sequential graph computations. Proc. VLDB Endow. **10**(12), 1889–1892 (2017)
18. Figueiredo, D.R., Ribeiro, L.F.R., Saverese, P.H.: struc2vec: learning node representations from structural identity. In: Proceedings of the 23rd ACM SIGKDD International Conference on Knowledge Discovery and Data Mining, Halifax, NS, Canada, pp. 13–17 (2017)
19. Getoor, L., Diehl, C.P.: Link mining: a survey. ACM SIGKDD Explor. Newslett. **7**(2), 3–12 (2005)
20. Gonzalez, J.E., Low, Y., Gu, H., Bickson, D., Guestrin, C.: PowerGraph: distributed graph-parallel computation on natural graphs. In: 10th {USENIX} Symposium on Operating Systems Design and Implementation ({OSDI} 2012), pp. 17–30 (2012)
21. Gonzalez, J.E., Xin, R.S., Dave, A., Crankshaw, D., Franklin, M.J., Stoica, I.: GraphX: Graph processing in a distributed dataflow framework. In: 11th {USENIX} Symposium on Operating Systems design and Implementation ({OSDI} 2014), pp. 599–613 (2014)
22. Granville, K.: Facebook and Cambridge Analytica: what you need to know as fallout widens. New York Times **19**, 18 (2018)
23. Grover, A., Leskovec, J.: node2vec: scalable feature learning for networks. In: Proceedings of the 22nd ACM SIGKDD International Conference on Knowledge Discovery and Data Mining, pp. 855–864 (2016)
24. Inc, A.: Euler: a distributed graph deep learning framework (2019). https://github.com/alibaba/euler
25. Inc, A.: Graph Compute (2020). https://help.aliyun.com/document_detail/134189.html
26. Inc, A.: Amazon Neptune (2018). https://aws.amazon.com/neptune/
27. Jones, N.D.: An introduction to partial evaluation. ACM Comput. Surv. (CSUR) **28**(3), 480–503 (1996)
28. Khan, B.S., Niazi, M.A.: Network community detection: a review and visual survey. arXiv preprint arXiv:1708.00977 (2017)

29. Khayyat, Z., Awara, K., Alonazi, A., Jamjoom, H., Williams, D., Kalnis, P.: Mizan: a system for dynamic load balancing in large-scale graph processing. In: Proceedings of the 8th ACM European Conference on Computer Systems, pp. 169–182 (2013)
30. Leskovec, J.: Graph representation learning with graph convolutional networks
31. Malewicz, G., et al.: Pregel: a system for large-scale graph processing. In: Proceedings of the 2010 ACM SIGMOD International Conference on Management of Data, pp. 135–146 (2010)
32. McSherry, F., Isard, M., Murray, D.G.: Scalability! But at what {COST}? In: 15th Workshop on Hot Topics in Operating Systems (HotOS {XV}) (2015)
33. Neo4j, I.: Neo4j Graph Platform (2021). https://neo4j.com/
34. Nielsen, A.: Nielsen global online consumer survey: trust, value and engagement in advertising. Ad week Media (2010)
35. Pearl, J.: Theoretical impediments to machine learning with seven sparks from the causal revolution. arXiv preprint arXiv:1801.04016 (2018)
36. Perozzi, B., Al-Rfou, R., Skiena, S.: DeepWalk: online learning of social representations. In: Proceedings of the 20th ACM SIGKDD International Conference on Knowledge Discovery and Data Mining pp. 701–710 (2014)
37. Ramalingam, G., Reps, T.: On the computational complexity of dynamic graph problems. Theoret. Comput. Sci. **158**(1–2), 233–277 (1996)
38. Salihoglu, S., Widom, J.: GPS: a graph processing system. In: Proceedings of the 25th International Conference on Scientific and Statistical Database Management, pp. 1–12 (2013)
39. Tian, Y., Balmin, A., Corsten, S.A., Tatikonda, S., McPherson, J.: From "think like a vertex" to "think like a graph". Proc. VLDB Endow. **7**(3), 193–204 (2013)
40. Valiant, L.G.: A bridging model for parallel computation. Commun. ACM **33**(8), 103–111 (1990)
41. Wu, Z., Pan, S., Chen, F., Long, G., Zhang, C., Philip, S.Y.: A comprehensive survey on graph neural networks. IEEE Trans. Neural Netw. Learn. Syst. **32**, 4–24 (2020)
42. Yan, D., Cheng, J., Lu, Y., Ng, W.: Blogel: a block-centric framework for distributed computation on real-world graphs. Proc. VLDB Endow. **7**(14), 1981–1992 (2014)
43. Zhu, R., et al.: AliGraph: a comprehensive graph neural network platform. arXiv preprint arXiv:1902.08730 (2019)

Design of Southbound Interfaces in Heterogeneous Software-Defined Satellite Networks

Xiupu Lang and Lin Gui[✉]

Shanghai Jiao Tong University, 800 Dongchuan Road, Shanghai 200240, China
{langxp2018,guilin}@sjtu.edu.cn

Abstract. In recent years, software defined networking (SDN) has been a promising tool for achieving flexible satellite networks, exploiting the advantage that it follows the idea of "renewing switch functions without changing hardware". Southbound interfaces (SBIs) are recognized as pivotal technologies in SDN for flexible network management and control, though, are faced with tough challenges posed by satellite networks. The OpenFlow, a representative and pervasive protocol in existing SBIs, only applies to wired communication systems such as Ethernet. However, satellite networks have unique characteristics in structure and service types distinguished from terrestrial wired networks, which means the unmodified implementation of existing SBIs cannot satisfy the needs of satellite networks. Based on the problem above, this paper aims to redesign a SBI well fitting to heterogeneous software defined satellite networks. Firstly, this paper analyzes the applicability of existing SBIs in a heterogeneous scenario comprising remote sensing satellites and relay satellites. Secondly, a SBI design framework is proposed, in which a scheme of building an intermediate layer on top of the OpenFlow is presented for reducing the protocol complexity. This framework gives a paradigm of the OpenFlow extension and serves as a exploratory work for achieving unified resource management in software-defined satellite networks.

Keywords: Heterogeneous space information networks · Software defined networking · Southbound interface extension

1 Introduction

With the development of human space technology, human's life has gradually become closely connected to space information. Constructing satellite-centered systems integrating various civilian and military applications, combined with the vigorous development of space information networks is expected to be an important direction [2,10,17]. However, past satellite systems organized by function-customized satellites are only able to serve a specific group of users. Besides, the lack of a multi-user and multi-task management mechanism incurs the tough

© Springer Nature Singapore Pte Ltd. 2021
Q. Yu (Ed.): SINC 2020, CCIS 1353, pp. 179–189, 2021.
https://doi.org/10.1007/978-981-16-1967-0_13

problem of resource sharing and application coordination [19]. To date, how to efficiently design management and control protocols for heterogeneous satellite network to realize information sharing and application collaboration has been a difficult problem in academia and the reasons are as follows. 1) Heterogeneity between different satellite systems. For example, there are great distinctions between UAVs (unmanned aerial vehicles) systems in near space and satellite systems in deep space with regard to movement patterns, resource attributes and service types. One single protocol cannot fit to all networks. 2) High dynamic in satellite networks. The time-varying characteristic of the network topology conduces no stable end-to-end links. Furthermore, protocols for time-varying networks are recognized as calculation-consumption and hard to reduce complexity as well. 3) Function-specialized payloads. The onboard payloads are designed to achieve a specific task since manufactured. Thus, the quality and standard of data respectively generated by satellite systems are not uniform, which poses great challenges to the design of cross-network resource management protocols. 4) The difficulty in maintenance and function upgrade for satellites in orbit. Onboard payloads cannot be modified once the satellite is launched. In response to inflexibility, resource management and control protocols need to be reconfigurable. Undoubtedly, satellite networks lack effective means of resource sharing and demand for better flexibility. Novel network architecture and network management and control methods are anxiously yearned improve the sharing and reconfiguration capabilities of satellite networks.

Software defined networking (SDN), a new network architecture that separates the control and data forwarding plane, has aroused great interest in applying SDN to spatial information networks with its openness, flexibility and programmability [6]. Applications of SDN into satellite networks have many notable advantages. 1) The unified access and management of heterogeneous networks can be attained through customized southbound interfaces (SBIs). Furthermore, the management of satellite resources under different granularity can be accomplished by redesigning actions of SBIs. 2) The operation of distributed routing protocols on each satellite node is not necessitated under the circumstance of the SDN's centralized control. In the meantime, central SDN controllers are able to optimize cross-network resource allocation under the global network view [5]. 3) The separation of control and forwarding planes, and the programmability in the control plane enable fast deployment of network function. To the best of our knowledge, many researchers have investigated the software-defined spatial information network and achieved a series of outcomes. In the aspect of routing strategy design, authors in [18] proposed a hybrid of centralized and distributed routing strategy based on SDN, aiming to optimize the delay of data transmission between quantum satellites. In terms of the placement of SDN controllers, the work in [11] adapted to dynamic changes and reduced the establishment time of service flows in large-scale low-orbit satellite constellations, by means of dynamically choosing satellite controllers. The work in [1] introduced the concept of SDN/NFV in the gateway cluster to maximize the bandwidth of the fronthaul link regarding the strategy of gateway selection. With regard to the reliability optimization of SDN-based control links, authors in [3] improved the reliability of control links by jointly optimizing the power allocation of the gateways and the access scheme of control links. However, the aforementioned literatures mainly focus on the network control strategy based on SDN framework. Researches on the implementation of SDN into satellite networks, that is, researches on SBIs' adaption, are severely lacking. The work in [8] studies the signaling system of the satellite-ground hybrid network proposed a signaling scheme adapting to the dynamic environment. Authors in [4] applied the idea of multi-granularity hierarchical label proposed in [8] to extensions of the OpenFlow, and took advantages of the multi-flowtable provided in the OpenFlow to achieve multi-granular control of satellite resources. However, the work in [8] only conducts a theoretical analysis and does not design a practical signaling scheme which can be used in specific satellite scenarios. Shortcomings in literature [4] lies in problems of high protocol complexity and low protocol efficiency, though, the existing OpenFlow is extended according to specific satellite scenarios.

This paper is committed to analyzing the applicability and redesigning a SBI for the management and control in heterogeneous satellite networks. The main contribution of this paper is to propose a layered architecture of the OpenFlow extension to realize real-time monitoring of network performance, and to improve network flexibility and reconfigurablity. The remaining parts of this article are organized as follow. Section 2 takes a heterogeneous network composed of sensing/relay hybrid satellites as an example to analyze the applicability of the OpenFlow. Section 3 first demonstrates the SBI design norms featuring layered structure, and then presents a paradigm of achieving real-time monitoring for dynamic heterogeneous satellite networks based on the OpenFlow. Section 4 summarizes the whole work and gives a conclusion.

2   Problem Analysis

The OpenFlow, one of the widely-used SBIs standardized by ONF (Open Network Foundation), is responsible for signaling interaction between SDN controllers and SDN-supported switches. However, the OpenFlow is designed for management and control for wired networks such as Ethernet. For example, OpenFlow1.0, the first version of the standardized SBI, features the 12-tuple that forms the header field of the flow table. This 12-tuple represents L2-L4 protocol fields of Ethernet, including physical port number, MAC (Media Access Control) address, IP address, VLan ID and etc. Thus, the OpenFlow can only process IP data packets [9], which is only one aspect of the limitations of applying the OpenFlow into satellite networks. This section below will take a specific satellite scenario as an example to systematically analyze the applicability of the OpenFlow in a satellite network.

2.1   Applicability Analysis Towards Hybrid Satellite Networks

The hybrid satellite scenario is composed of 3 low-orbit remote sensing satellites and 3 relay satellites orbited in geostationary position, as shown in Fig. 1.

Assuming that the hybrid heterogeneous network adopts the SDN architecture and is used to complete remote sensing tasks, the OpenFlow should meet the functional requirements of the remote sensing observation task during all the process from task generation stage to task execution stage. The workflow of the hybrid network to perform remote sensing tasks is as follows.

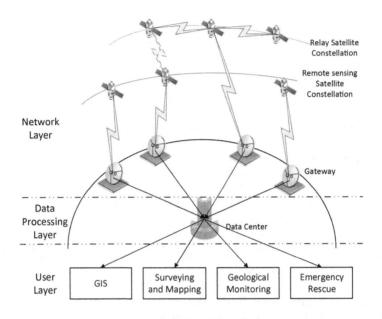

Fig. 1. A scenario of hybrid satellite constellations.

Information Acquisition and Maintenance Stage. When there is no task arrival, each controllable satellite need to set up control channel with the central controller (CC, CC lies in the data center shown in Fig. 1). After control messages are successfully exchanged, the CC needs to maintain keep-alive information (e.g., satellite position and battery remaining energy), resource attributions including the number of communications/remote sensing payloads and their parameters, as well as the real-time status of resource occupancy.

Mission Planning Stage. When receiving a task request of remote sensing, the CC needs to obtain the knowledge of observing capabilities from each remote sensing satellites in real time, such as satellite A is equipped with hyper-spectral radar while B is equipped with SAR radar. Then, considering mission requirements, a single or a group of satellite are arranged by CC to perform the remote sensing mission, and the strategy for transferring the data stream is dispatched to the corresponding satellite node using flowtables.

Mission Execution Stage. Observation data starts to flow into the remote sensing satellite once mission is being executed. Data is regarded as a "network stream" by SDN, and its forwarding direction is decided by matched flow entries. Owing to the fact that there is possibility no gateway is visible, the data stream tend to remain stored inside the switching satellite or be forwarded to relay satellites via microwave or laser links. When it comes to data transmission between relay satellites and ground, it is vital to select appropriate frequency and access scheme to prevent signal interference.

From the perspective of task execution, it is worth noted that the high dynamic and wireless communication in satellite network make it quite different from wired networks. The existing SBIs cannot meet the functional requirement of satellite workflow, which is explained below.

1) Overheads induced by controllers are too heavy. In the SDN network, all the forwarding strategies of the switch are dispatched and updated by central controllers, which induces a huge overhead. There are many studies on reducing overhead in the terrestrial network, such as [12,13]. Due to the satellite mobility, the number of satellite under the control will dynamically change, and the overhead problem will become more severe.

2) Unique service types and resource attributions. Compared with the Ethernet, satellite systems have its inherent TT&C (Telemetry, track and Command) services [15], which creates new service requirements. In addition, satellites, different from power-supported switches, are resource-limited systems. Therefore, the Openflow is forced to add additional statistical information, including the remaining battery power of the satellite, the number of idle communication devices and other system constraints.

3) Multi-granularity resource management with time constraints. In Ethernet, all SDN switches are standardized characterized by port number, throughput and etc. and network topology is fixed. Thus, the CC can uniformly schedule data traffic distributed on each link. However, resources in satellite network are complex and hard to uniformly describe. Besides, the trait of time-varying and delay-nondeterministic network make the flowtable valid only if it successfully reaches its destination node before network changes. Thus, the management ability with multiple granularity is required for SBIs.

4) Needs for new actions in protocols. Due to the lack of end-to-end reliable links, a store-and-forward communication strategy (e.g. the bundle protocol [14]) is often applied in satellite scenarios. This requires SBIs to support the actions of the bundle protocol. As the satellite transmission protocol matures, the SBI will be confronted with support for more new actions.

5) Huge distinctions between wireless and wired communication in MAC. One of the essential functions for MAC in Ethernet is to realize topology discovery and logically form a tree through LLDP (Link Layer Discovery Protocol) [7]. Due to the limitation of transmission power, directional antennas has been widely used, which causes the satellite network to not only perform topology discovery in MAC, but also perform real-time frequency selection and link allocation [16].

Apart from the above reasons, small-scale satellite constellations are unable or unnecessary to implement the whole network functions of Ethernet due to their limited on-board processing capabilities. Therefore, there exists redundancy in the OpenFlow design in regard with satellite networks. In summary, the Open-Flow cannot adapt to the dynamics of satellite scenarios, resource heterogeneity, and the characteristics of wireless communication, and cannot achieve unified management and control of satellite resources with multi-granularity.

3   The Extension of the OpenFlow for Heterogeneous Satellite Networks

3.1   Framework for the OpenFlow Extension

Based on the analysis in Sect. 2, we can derive that the OpenFlow cannot be applied into wireless and dynamic scenarios, and cannot conduct actions such as link and wavelength allocation. In order to solve the above problems, the most intuitive way to extend the SBI is to add resource labels existing in satellite networks into the OpenFlow's header filed. Since the OpenFlow1.2, it has canceled the fixed-length matching field and replaced it with OXM (OpenFlow eXtensible Match) with a TLV structure. The protocol user can write any number of OXM TLVs into the structure of ofp_match to realize the expansion of header fields. Therefore, we can write all kinds of resource attributions, microwave/laser port information, and frequency information into the header field of the OpenFlow to realize the control of heterogeneous satellite networks. However, the solution above have the defect of high complexity in the protocol design. Satellites have not been standardized, thus, load types and resource capabilities possessed by various satellites are not the same. If information for all kinds of satellite is added to the header field, it will be very large and the match speed of flow table slows down, causing the decrease in switch throughput. Besides, There might be cases where the header field proposed by one satellite system is invalid to that of another satellite systems.

To tackle with the problem of high complexity, the complexity of OpenFlow is moved to the upper level in this paper, and a method of building an intermediate layer on top of the protocol is proposed to realize the adaptation of the SBI, and the management and control of the heterogeneous satellite network at a minimum cost. This paper takes into account that there are both commonalities and characteristics between different types of satellites. The common parts among satellites are the main and usual parts for network decision makers, such as the satellite's idle communication bandwidth, remaining battery capacity, available storage space, etc. The characteristics of satellites are reflected in the specific resource requirements for specific tasks. For example, remote sensing tasks for real-time observation of a certain area require to identify the time window of the observation, or a certain communication task demands for laser communication and requires a reasonable allocation of wavelength. When there is no task with specific resource requirements, the obtainment of characteristic information is unnecessary, causing the deletion of characteristic information from the header field. The relationship between commonalities and characteristics resembles the relationship between the parent and the child. Part of the characteristics can be calculated from the commonalities. Figure 2 gives two examples of the relationship between the two types of information. The position information of a single satellite can be described by the Kepler's six elements. The position information and the energy consumption model, combined with the remaining energy at a certain time can constitute a "general satellite" model. Assuming that there is a remote sensing task, the attribute parameters of the above three commonalities can calculate the time window within which the satellite can observe the targeted area. In another example describing the satellite communication capabilities, we can use transmission bandwidth and the number of transponders to define the communication capabilities of general satellites. When the network needs to make use of one specific microwave/laser communication scheme, we can use the conversion to generate specific portal types. It can be learned that we can acquire and maintain the commonalities on a regular basis, and when there is a need for specific tasks, we can obtain the characteristics like a "on-off" switch.

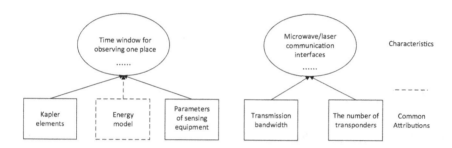

Fig. 2. The relationship between characteristics and commonalities.

Based on the ideas above, this paper proposes a SBI design framework, as shown in Fig. 3. The SBI is composed of THE extended OpenFlow and an intermediate layer above it. The extended OpenFlow treats any type of satellite as a general satellite, maintains the common information in real time according to the general satellite model, and reports the common information of the satellite to the middle layer. What the middle layer maintains is commonality/characteristic information, and it will decide whether to convert the commonality information into specific characteristic information and send it to the CC according to the needs of tasks. The advantage of the above architecture is that the complexity of the OpenFlow is low, and the access of heterogeneous satellite networks can be achieved without modifying the OpenFlow. Adaptations towards heterogeneous satellite networks is mainly completed in the middle layer. The feature of this part is on-demand, that is, common information will be further transformed into

characteristic information required for task execution and delivered to the CC
only when resource scheduling should be performed at a certain granularity or
cross-network resource collaboration is needed.

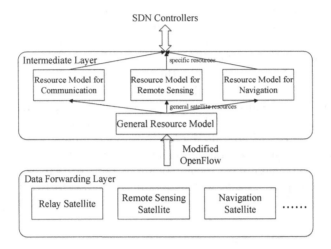

Fig. 3. A framework for SBIs design.

3.2   The Extension of the OpenFlow Ased for General Satellite Model

This section gives a paradigm of the OpenFlow extension, iin order to meet
the management and control requirements of heterogeneous satellite networks.
Although the design ideas described in Sect. 3.1 reduce the complexity of protocols, more attributions originating from general satellite model are needed to
add in the header field in the OpenFlow. First, it is assumed that the resource
state, the satellite position and time label together constitute a general satellite model. And the resource state can be categorized into these three aspects of
communication, calculation, and storage. Since the satellite's maximum resource
capacity and the location at any time can be stored in the CC in advance, real-
time information of satellite resources is more valuable and should be acquired
through the OpenFlow. OpenFlow1.3 and later versions use the "experiment"
type to implement user-defined controller management messages. This paper
extends the Experimenter Multipart message to obtain real-time resource status
information of the general satellite model in real time and the code is shown
below.

```
struct ofp_experimenter_multipart_request {
uint32_t experimenter;
uint32_t exp_type;
};

struct ofp_experimenter_multipart_reply{
uint32_t experimenter;
uint32_t exp_type;
uint32_t cpu_freq;
/* Available computing power,
remaining CPU frequency, GHz */

uint32_t storage;
/* Available storage capacity,
remaining storage capacity, MB */

uint32_t num_transponder;
/* Available communication capacity,
the number of idle transponders */

uint32_t bandwidth[num_transponder];
/* Available communication capacity,
idle bandwidth, Mbps */

uint32_t battery_capacity;
/* remian battery capacity, KJ */
};
```

4   Conclusion

This paper systematically analyzes the applicability of the OpenFlow in heterogeneous satellite networks, and concludes that the existing OpenFlow has obvious defects in terms of protocol overhead, applicable scenarios, and protocol functions. Aiming at the significant problem of protocol inefficiency caused by directly adding satellite resource information to the protocol header field, this paper proposes a layered framework for SBIs design, in which we build an intermediate layer on top of the existing OpenFlow. The intermediate layer realizes the information exchange of coarse-grained general satellite model and the fine-grained one. it reduces the complexity of the OpenFlow for the function extension, and the on-demand working mode of the intermediate layer significantly improves the protocol efficiency. Besides, the layered architecture makes it unnecessary to change the internal implementation of the OpenFlow when a heterogeneous satellite network is accessed. Finally, based on the proposed design ideas, this paper gives a paradigm of the OpenFlow extension, which serves as a exploratory work for achieving unified resource management in software-defined satellite networks.

References

1. Ahmed, T., Ferrus, R., Fedrizzi, R., Sallent, O., Gelard, P.: Satellite gateway diversity in SDN/NFV-enabled satellite ground segment systems. In: IEEE International Conference on Communications Workshops (2017)
2. National Natural Science Foundation of China: 2016 project guide for the research plan of basic theory and key technology of spatial information network. [EB/OL]. http://www.nsfc.gov.cn/publish/portal0/tab38/info51946.htm. Accessed 25 March 2016
3. Cho, W., Choi, J.P.: Cross layer optimization of wireless control links in the software-defined LEO satellite network. IEEE Access **7**, 113534–113547 (2019)
4. Feng, M., Xu, Z., Wang, C., Zhang, Y., Xiao, Y.: SDN-based satellite network and the extension of southbound interface protocol. Radio Communication Technology **05**(43;259), 23–27 (2017)
5. Fu, H.: Research on SDN-based resource pricing algorithm and reliability transmission strategy. Ph.D. thesis
6. Jammal, M., Singh, T., Shami, A., Asal, R., Li, Y.: Software-defined networking: state of the art and research challenges. Comput. Netw. **72**, 74–98 (2014)
7. Li, Y., Cai, Z.P., Xu, H.: LLMP: exploiting LLDP for latency measurement in software-defined data center networks. J. Comput. Sci. Technol. English Edn. **33**(2), 277–285 (2018)
8. Ma, F.: Research on the extension of signaling protocol and its prototype implementation in satellite-ground hybrid networks. Ph.D. thesis (2016)
9. McKeown, N., et al.: OpenFlow: enabling innovation in campus networks. ACM SIGCOMM Comput. Commun. Rev. **38**(2), 69–74 (2008)
10. Min, S.: Discussion on space-based comprehensive information network. Inter. Aerosp. **8**, 46–54 (2013)
11. Papa, A., Cola, T.D., Kellerer, W., Machuca, C.M., Vizarreta, P.: Dynamic SDN controller placement in a LEO constellation satellite network. In: IEEE Global Communications Conference: Selected Areas in Communications: Satellite and Space Communications (Globecom2018 SAC SSC) (2018)
12. Pranata, A.A., Jun, T.S., Kim, D.S.: Overhead reduction scheme for SDN-based data center networks. Comput. Stan. Interfaces **63**, 1–15 (2019)
13. Qiang, H., Min, W., Huang, M.: Openflow-based low-overhead and high-accuracy SDN measurement framework. Eur. Trans. Telecommun. **29**(2), 101–117 (2018)
14. Sabbagh, A., Wang, R., Zhao, K., Bian, D.: Bundle protocol over highly asymmetric deep-space channels. IEEE Trans. Wirel. Commun. **16**, 2478–2489 (2017)
15. Thiruvoth, D.V., Raj, A.B., Kumar, B.P., Kumar, V.S., Gupta, R.D.: Dual-band shared-aperture reflectarray antenna element at Ku-band for the TT&C application of a geostationary satellite. In: 4th International Conference on Recent Trends on Electronics, Information, Communication & Technology (RTEICT), pp. 361–364. IEEE (2019)
16. Ullah, M.A., Alam, T., Misran, N.B., Islam, M.T.: A 3D directional antenna for s band small satellite communication system. In: 2017 6th International Conference on Electrical Engineering and Informatics (ICEEI) (2017)

17. Wang, H., Han, Z., Fu, J.: Development of space-earth integrated information technology for promoting the upgrading and transformation of satellite application industry. Appl. Satellites **1**, 30–34 (2014)
18. Wang, Y., Zhao, Y., Chen, W., Dong, K., Zhang, J.: Routing and key resource allocation in sdn-based quantum satellite networks. In: 2020 International Wireless Communications and Mobile Computing (IWCMC) (2020)
19. Wu, M., Wu, W., Zhou, B., Lu, Z., Zhang, P., Qin, Z.: Visions of the overall structure of the space-earth integrated information network. Satellite and Networks (2016)

Author Index

Printed in the United States
by Baker & Taylor Publisher Services